Renewable Hydropower Technologies

Renewable Hydropower Technologies

Vierah Hulley

ARCLER
PRESS

www.arclerpress.com

Renewable Hydropower Technologies

Vierah Hulley

Arcler Press
224 Shoreacres Road
Burlington, ON L7L 2H2
Canada
www.arclerpress.com
Email: orders@arclereducation.com

ISBN: 978-1-77469-139-7 (Hardcover)

ABOUT THE AUTHOR

Vierah Hulley is an internationally experienced expert in Environmental and Earth Sciences and Management. She has an extensive vocational background in the areas of environmental policy development, environmental risk management, natural resources management, contaminated site remediation, sustainable development, and spatial analytics. She is the founder and Managing Director of HL Nexus: The Data Imaginarium; a company dedicated to optimizing natural resources management with data science. Vierah holds a master's degree in Geology and a PhD in Geohydrology, with a focus on environmental and spatial science.

TABLE OF CONTENTS

LIST OF FIGURES

LIST OF ABBREVIATIONS

ABT	Availability Based Tariff
AFC	Annual Fixed Charges
AHP	Analytical Hierarchy Process
BBA	Biochimica et Biophysica Acta
BHEP	Budhil Hydro Electric Project, India
CAES	Compressed Air Energy Storage
CDM	Clean Development Mechanism
CEA	Central Electricity Authority
CRU	Catalytic Reforming Units
EIA	Environmental Impact Assessment
EIR	Environmental Impact Report
EPR	Energy Payback Ratio
EU	European Union
FC	Fixed Charge
GHG	Global Greenhouse Gas
GMBD	Gazi Mühendislik Bilimleri Dergisi
IDGTE	Institution of Diesel and Gas Turbine Engineers
IJITEE	Innovative Technology and Exploring Engineering
IL	Installation License
IRENA	International Renewable Energy Agency
ISSCC	International Solid-State Circuits Conference
KWs	Kilowatts
LCA	Life-Cycle Assessment
MWs	Megawatts
ORC	Organic Rankine Cycle
OTEC	Ocean Thermal Energy Conversion
PAT	Pump as Turbine

PEDG	Power Electronics for Distributed Generation Systems
PH	PowerHouse
PHP	Pico Hydropower
PL	Preliminary License
PRO	Pressureretarded-Osmosis
PSU	Public Sector Undertakings
REC	Renewable Energy Certificates
SEB	State Electricity Boards
SHP	Small Hydropower Plant
SHP	Specialists in an Project
SHP	Small Hydropower
TWh	Terawatt-Hours
UNFCCC	United Nations guidelines Core Convention on Climate Change
UNIDO	United Nations Industrial Development Organization
WCD	World Commission on Dams

PREFACE

Water is considered the fundamental component of life on earth. Moreover, it is also globally exploited as a sustainable energy source. The technology of power generation utilizing water has been established and has matured enough over time to offer an economic solution to harnessing hydro power and tidal power. Water is a renewable source of energy with minimum impacts on the environment. Because of the broad variation in settings of power generation plants, the operation and technical design of hydropower plants differ extensively. Operations and technical designs of power plants are usually governed by legal requirements and local practices. The average power generation capability of hydropower systems varies between a few kilowatts (KWs) and thousands of megawatts (MWs), and plant sites are usually found near water streams and rivers.

This book elucidates different categories of hydropower systems. The book is designed in such a way that it creates harmony between shorter book chapters and simultaneously presenting a comprehensive text on hydropower generation for the readers.

The book is divided into eight chapters. Each chapter of the book explains the fundamental concepts regarding a particular topic. Chapter 1 offers a detailed introduction of renewable hydropower and hydropower plants along with a generic overview of the global standing of hydropower. Chapter 2 familiarizes the readers with the fundamentals of small hydropower technology with emphasis on costing and materials selection. In Chapter 3, the fundamental terminologies and legal requirements of hydropower plants are discussed in detail. Chapter 3 also contains information about the clean development mechanism with specific examples of India. Chapter 4 discusses the technical and physical fundamentals of hydropower plants. A detailed analysis of fluid mechanics and the sustainability of plants is also discussed in the chapter.

Chapter 5 thoroughly discusses the management models and economics of hydropower technologies. Every hydropower plant contains some significant components (structural, electrical, auxiliary, etc.) that are essential for functioning components of hydropower systems. Chapter 6 provides a detailed discussion about various components of a power plant.

Hydropower is often termed "green energy" because its production does not generate harmful emissions. However, the main environmental risk remains the exploitation of hydropower sites that are created by destroying the natural ecosystem. Hydropower plants play a major role in combating global warming and providing sustainable green solutions for the world. Chapter 7 focuses on the environmental impacts of hydropower systems. Finally, Chapter 8 provides information about hydraulic turbines and their potential to harness ocean energies. The chapter also contains a detailed discussion about the operational aspects of turbines and their applications in hydropower.

This book is intended for a generic audience from different fields, including electrical, mechanical, and civil engineering. This book is aimed at presenting a good preliminary background for professionals from interdisciplinary fields. However, students in the field of renewable energy and hydropower systems can also benefit from the comprehensive chapters of the book.

—Author

Chapter 1

Introduction to Renewable Hydropower

CONTENTS

1.1. INTRODUCTION

Human life in these is directly or indirectly reliant on one of the forms of energy. The system's capability to perform work is named energy. Energy can be transformed in three ways from one form to another (Georgescu, 1975):

- as in a belt drive, by performing mechanical work;
- by electromagnetic fields, such as in an electromotor;
- by exchange of heat, such as in a steam engine.

Physically, energy can be transformed from one form to other. The electrical and thermal energies are of particular significance. Gas, oil, and electricity are the main forms of energy being used nowadays. There are various forms of energies such as (Dincer & Cengel, 2001; Kalogirou, 2004):

- kinetic energy: the energy present due to motion of a body which is same as the square of its speed times one-half the mass of the body.
- potential energy: energy possessed because of the position within a physical system (variation in altitude).
- chemical energy: the energy held in a body due to bonding of atoms and different other types of agglomerates of matter.
- thermal energy: the energy present due to the temperature variations in a body.
- nuclear energy: the energy discharged through a nuclear reaction because of fusion or fission. Also, it is named atomic energy.
- electrical energy: the energy through a conductor due to the movement of electric charge.

Nowadays, the production of heat and thermal energy are the major forms of energy that meet the demands of the major share of the world's energy necessities, acquired through the burning of fossil fuels (Ayres et al., 2003). Dependency on fossil fuels is inevitable considering the requirements of energy; however, biomass plays a small part to meet some of the needs of the world's energy. Heat energy produced has a limitation that it can be utilized without first conversion to some useful form before generating mechanical work i.e., moving a car or running a pump. In the same way, it has no direct use in operating a computer or lighting an electric lamp. It is termed as 'low grade' energy because of the above-mentioned reasons. Contrary to this, kinetic energy, in many cases, commonly is an ideal type of energy when it is to be used in the method of shaft rotation. Only a part of heat energy can be transformed into useful mechanical work according to the laws of thermodynamics. The second law of thermodynamics is based on law of energy conversion. Considering the second law of thermodynamics, a trend is observed that the heat energy converted to the mechanical work is higher when heat is being delivered is higher at higher temperature, and similarly, the temperature is lower at which the remaining heat is put into the atmosphere (Dincer & Cengel, 2001). Moreover,, materials properties play an important role in determining the maximum temperature which is normally limited by the material's heat bearing properties, while, the temperature of the environment is typically lesser than the lower temperature of heat sink. The residual part of heat is released into the environment as waste heat, which is not transformed into mechanical work due to the above limitation. This limitation is due to the reason that the efficacy of oil, gas, and coal power plants does not surpass 50% (Brown & Ulgiati, 2004; Arutyunov & Lisichkin, 2017).

Electrical and mechanical energy are preferred over thermal energy and are termed as 'high-grade energy' since their conversion to all other forms of energy is feasible with minor losses. Electrical energy, because of its wide distance transportation through transmission losses without much losses, is preferred the most, which is relatively difficult for mechanical energy (Fazeli et al., 2016; Smith et al., 2020).

Despite all these facts, the law of conversion of energy is not applicable for any of the forms of energy that are produced or consumed, but it is only applicable for the conversion of energy from one form to another (Suberu et al., 2013). The terms "energy consumption" and "energy generation" are usually used in the energy industry and daily life. Economically, the relationship between consumers and producers is involved. The energy which has been "consumed" is economically worthless form, and the energy generated is able to do useful work. Therefore, in this book, these terms will be used in the same way (Warr et al., 2010).

1.2. ENERGY UNITS

Energy technology and the energy industry usually utilize various energy units, that creates difficulty when comparing data based on energy requirements, energy consumption, and kinds of sources of energy used (19). For this purpose, Tables 1.1 and 1.2 cover a list of commonly used units, conversion factors, and prefixes. The kilowatt-hour (kWh) and the

Joule (J) are mandatory units of energy derived under the international unit system (SI) familiarized in 1960. Watt is the unit used in the SI unit system for power. The cars power is presented in kilowatts (kW or 1,000 watts) while that of bulbs in Watts (W) and in megawatts for power stations (MW, or 1,000,000 watts).

Table 1.1. Energy Units Conversion

Unit	kJ	kWh
1 kilojoule (kJ)	–	0.000278
1-kilowatt hour (kWh)	3,600	–
1 MW annum	–	8,760,000

Table 1.2. Abbreviations and Prefixes

Prefix	Abbreviation	Exponent	Number
Kilo	k	10^3	Thousand
Mega	M	10^6	Million
Giga	G	10^9	Billion
Tera	T	10^{12}	Trillion
Peta	P	10^{15}	–
Exa	E	10^{18}	–

Domestic usage of electrical energy is represented in kilowatt-hours (kWh). Assume, an electric instrument consumes 1 kWh when worked for 1 hour having a capability of 1,000 W. In developed countries, a typical domestic home usage is between 6,000 and 3,000 kWh per year, whereas for developing nations this figure is comparatively lesser. Contrary to this, terawatt-hours (TWh) is the unit used for the electric energy production of a state which represents 1 billion kilowatt-hours, e.g., in the year 2010, Germany produces electric energy of almost 630 TWh (Toan et al., 2012; Ebrahimi et al., 2014).

1.3. NATURAL WATER CYCLE

The water cycle is a process that represents the water existing in the cosmos that goes into a cycle through various phases and forms as shown in Figure 1.1. Water is in the form of solid (ice) in some parts of the cycle; in other parts, it is in the form of liquid (rain) or gas (water vapors). The heat energy from the sun evaporates water from lakes, rivers, seas and also from the soil and the plants on the land. A process called "evaporation" turns this water into an unseen gas named water vapors. The water vapors, as they rise into the atmosphere, become cooler (Sun et al., 2016; Zeng et al., 2016).

In going high, some of the water vapors convert into droplets of water through a process named "condensation" since the moisture-holding capacity the warm air is much higher than that of the cool air. The tiny water droplets form clouds in the sky. Due to their weight, larger droplets are formed by the combination of these smaller droplets, they fall to the earth as snow, hail, or rain. A large portion of the water flows through the rivers and streams into the sea after falling on land. Some stay as ice and some get soaked into the ground. Eventually, the water cycle repeats its process after the water falls into the seas and rivers (Zhang et al., 2014).

The energy of water during this journey is used for the generation of power through the hydropower plants. In some sense, the hydropower plants may be regarded as 'man-made barriers' in the way of the water cycle after falling as snow or rain, before moving into the sea.

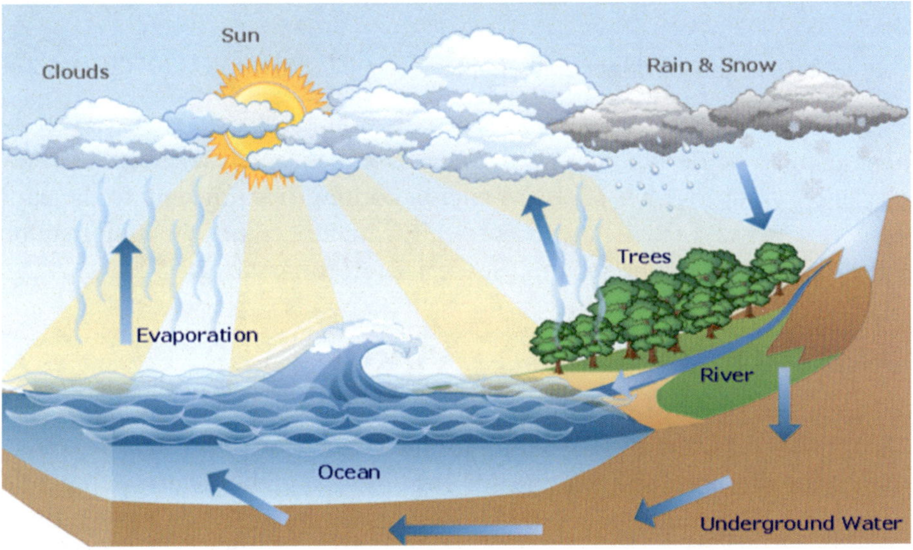

Figure 1.1. The water cycle .

[Source: https://www.vedantu.com/question-answer/draw-a-neat-and-labeled-diagram-of-water-cycle-class-11-biology-cbse-5f7be0874010be40a7d7d080.]

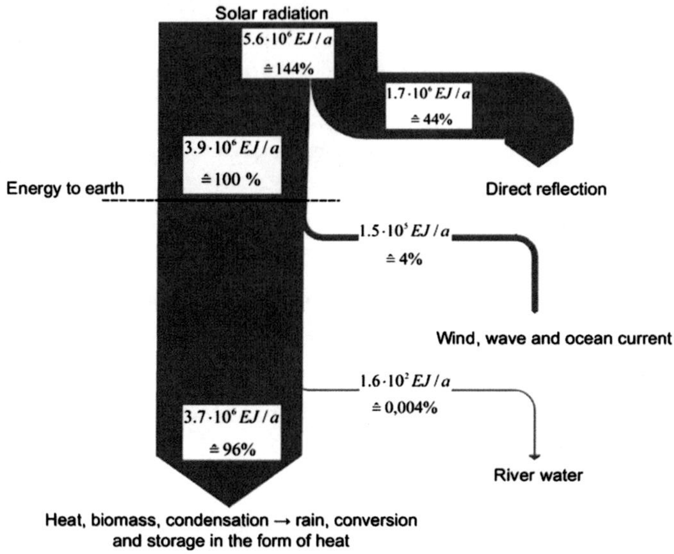

Figure 1.2. Energy flow diagram of solar radiation.

[Source: https://link.springer.com/chapter/10.1007/978–3-642–20709–9_1.]

All the forms of energy on earth, including hydro energy, can be considered to be derivative of solar energy except nuclear energy. 22% of solar energy out of 3.9 9 10^6 EJ of energy coming from the sun to the earth every year is use d for the creation of rain which becomes the major source of hydropower as can be seen in Figure 1.2 (Alkon et al., 2019).

1.4. INTRODUCTION TO HYDROPOWER

Since the inception of civilization, water power has been used by man. It was utilized as the major foundation for the generation of mechanical power along with the burning of wood for heating and light. There is potential energy present in water going to lower levels from higher due to its altitude that is transformed into kinetic energy during downhill movement. Water power is a combination of all these forms of energy. As it is transformed constantly in a normal way hence it is termed as a renewable source of energy (Şahin, 2004; Bilgen et al., 2008).

1.4.1. Classification of Hydropower Plants

Various aspects are considered in categorizing the hydropower plants. For example, on the baseis of the functioning of hydropower plants, classification can be done by their type of construction, by the source of water, or by turbine as is presented in Figure 1.3. Various aspects of hydropower plants can be easily understood with the help of this categorization and to comprehend how some categorizations are related with one another (Belessiotis & Delyannis, 2000).

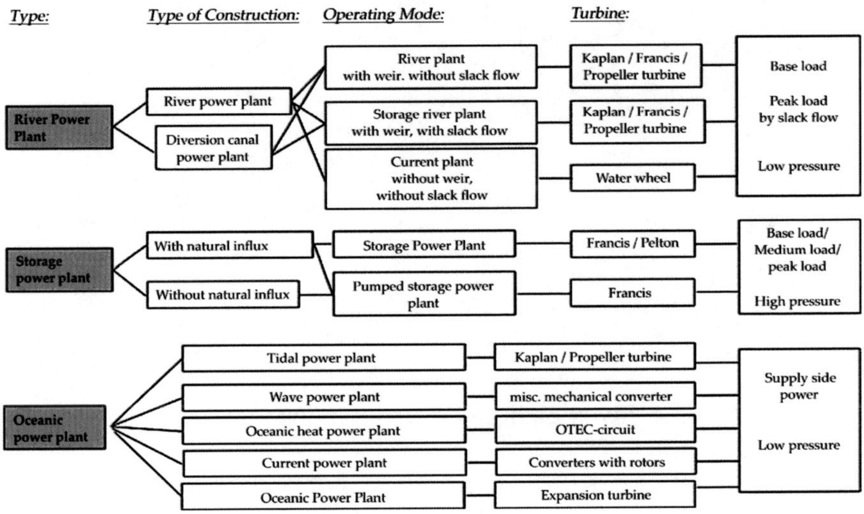

Figure 1.3. Overview of hydropower plants.

[Source: https://link.springer.com/chapter/10.1007/978–3-642–20709–9_1.]

1.4.1.1. River Power Plants

River power plants in some countries are also known as run-of-river power plants. Elevation drop of a river and the natural flow are used to produce electricity in this type of plant (Figure 1.4). a number of these power plants are served by a diversion canal, while others are served directly by a river as presented in Figure 1.5 (Yaseen et al., 2020). The former power plants are normally small in the capacity of power production. Diversion is used because the power plant is not served by the full capacity of the river but a small portion is used for feeding. There is a reduction in water feeding and subsequently power generation capacity of the power plant as represented in the figure. A canal or a diversion power plant compared with a power plant fed by the river cannot generate much electricity as a itself because of the limitation in the flow in the diversion canal.

Figure 1.4. The river power plant in Germany in wintertime.

[Source: https://link.springer.com/book/10.1007/978–3-642–20709–9.]

There are three main aspects in which the functioning mode of these plants can differ. First, the power plant fed with the river can work without a slack flow with a weir. The river stream is straight injected into the turbine without any stoppage by a dam (Hennig et al., 2013).

Operation with little storage is another mode of river power plant operation, which has a slack flow as well as a weir. A small reservoir is used for the storage of water. Water is stored for two main reasons: to get a definite depth of water to have the river maneuverable for ships and to serve as peak load by having water reserves. These reservoirs have a much smaller capacity equated to storage type reservoirs of hydropower plants (Andrade & Santos, 2015).

The last model which is applicable in minor river power plants lacks both slack flow and weir. Water wheel may not require slack flow or weir, as it is evident from the category "turbine" in Figure 1.3, as it goes just by the flowing water of a river. Weir is not present only in smaller power plants such as water wheel.

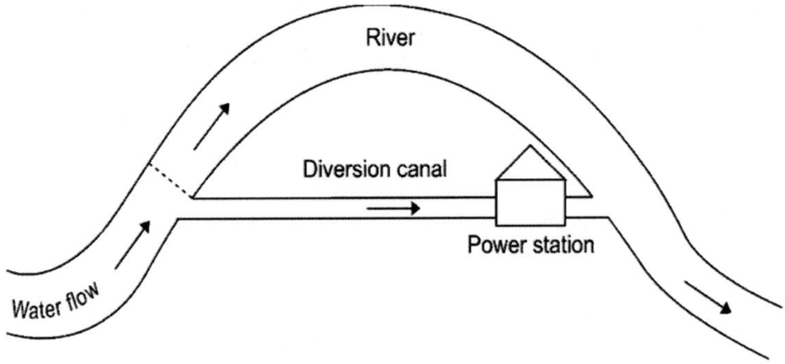

Figure 1.5. Scheme of diversion canal fed river power plant.

[Source: https://www.springer.com/gp/book/9783642207082.]

The turbines used are Propeller turbines, Kaplan turbines, and Francis turbines, for the first two functioning modes,. For huge masses of water and small height, these turbines are relatively suitable; bearing a relatively low pressure. At a height of about 10 m, the pressure of the water reaches up to 1 bar. There is a difference in pressure of about 0.6–0.7 bars with fall height of 6–7 m. Energy for the baseload is usually delivered through this kind of hydroelectric power plant. The range of capabilities of the river power plant may go to several hundred megawatts from a few kilowatts, dependent on the accessible height for the fall and volume of water (Richter & Thomas, 2007; Kubiak & Szczęch, 2021).

There are certain other advantages over storage power plants including prevention of relocation of people or animals near the river, and natural habitats are conserved, thus decreasing the impacts on environment. There is also one major disadvantage of river power plants that is a high dependency on the river run-off that can mismatch with the demand of the power (Storli & Lundström, 2019). Sometimes the only option is river power plants in conditions where e.g., river streams from one state to another and wherever stoppage of flow by one country is not allowed by agreement, depending upon security, irrigation, safety, or availability of drinking water in the other state.

1.4.2. Storage Power Plants

The storage power plant is another major type of hydropower plant (Figure 1.6). The natural influx of water may or may not be present in these types of power plants. A natural flux can be a pool of melted water draining from mountains or a rain-fed river (Storli & Lundström, 2019).

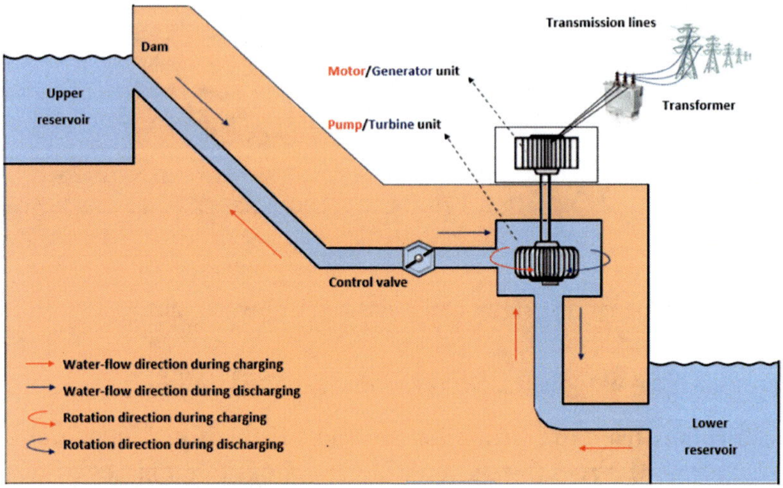

Figure 1.6. Scheme of storage power plant.

[https://www.researchgate.net/figure/Schematic-diagram-of-pumped-hydro-storage-plant_fig3_320755664.]

In different spots, bodies of water are linked through pipes between the mountains. This linkage is used to make the collection of water as efficiently as possible, which is potential energy e.g., to evade water draining that essentially means loss of energy.

The power plant works essentially as a pumped-storage power plant, when there is not sufficient influx naturally, the idea of that is clarified later in this part .

In storage-type plants, Pelton turbines and Francis turbines are used. For heights of fall of about 15–500 m, Francis turbines are used, while for heights of fall of up to 2,000 m, Pelton turbines are used. A large quantity of water flow can be managed by Francis turbines, and for relatively smaller flow Pelton turbines are utilized (Karhinen & Huuki, 2019). In storage power plants, Kaplan turbines are not usually used until heights of about 25 m. in subsequent chapters, the idea of turbine fittingness has been explained.

The pressure of water can go up to 80 bar when the water height in the mountains reaches to 800 m. A large amount of power can be generated by this huge quantity of water with this high pressure. Therefore, storage the potential energy of water is the function of the storage power plants is and permit its usage when required, particularly in peak and medium load hours, and also during the feeding rivers, to cater seasonal variations in water accessibility (Loots et al., 2015). As long as sufficient water is existing in the reservoir, the based load can be delivered. Storage power plants can store the meltwater in winter in the reservoir behind a dam until summer and can deliver a baseload for a few months. On the other hand, a normal pumped storage power plant cannot deliver a baseload, because of too small capacity that would function for only a few hours of baseload. For peak load hours, pumped stored water is mostly utilized.

1.4.3. Working of Pumped Storage Power Plants

From rivers and creeks in a comparatively huge catchment area around the reservoirs, storage power plants gather water in the storage medium which are situated at high altitude. Usually, not all of the water present in the watercourses is streamed to the reservoir for conservation reasons. There are two main reasons for the building of the reservoirs: first, these are used to meet the varying demand of electricity generated by the power plants; secondly, the potential energy of water is stored which serves as fuel for the hydropower plants at a lower level at any time. There are changing demands of electricity in a country; households and industrial plants do not require the same amount of power at every time (Andre, 1978). Cooking stoves function around at noontime and when required by the people in the evening. Similarly, industrial plants are shut down at night. In short, at certain times, there is additional consumption of electricity; therefore, power plants need to generate electrical output to meet these extra requirements. Breaks in major sporting events are evident examples of such peak demands. Consumption is low as long as the game is in progress, while during break time there is a surge in consumption due to lights, refrigerators, etc. This effect combined with a million houses burdens the electricity usage. The technical solution for the above issues is provided by the pumped storage hydropower plants-additional, quick power supply. The power production starts within 1–3 minutes as soon as the lid in front of the turbine is opened and the turbine starts turning. Gas turbines provide another solution for the production of quick additional power. These types of turbines can be seen in airplanes which produce quick power to aid lifting of the plan (Kubiak & Szczęch, 2021). Various types of storage power plants including pumped storage plants are utilized where there is a small little quantity of natural water available for use. Two reservoirs, one on the mountain and one in the valley are built in this structure, and water in the reservoirs is used to produce electricity. In case of a surge in the electricity demand, water is allowed to move to the lower basin from the upper basin through the turbine which in turn rotates to produce power. In case of decreased demand, a reverse process comes into action to draw water to the upper basin from the lower basin with the help of pumps that take power from the grid to make water available for future use. Herdecke on the Ruhr, Germany is the largest pumped storage plant which has been in operation since 1920 which works in the same way as explained earlier. Other pumped storage power plants in operation are located at Vianden, Luxemburg and the most modern pumped storage plants globally in Goldisthal, Thuringia, Germany (Hennig et al., 2013).

There are two main purposes of the pumped storage power plants. The most common and important use is the supply of electricity at peak demands and the other is the provision of reactive power. Consumers do not need just electrical power, but for physical reasons, there were two specific forms of energy required by the consumers: first if the active power or effective power, it is a power needed to drive a motor or to produce heat for the stove. To make the reactive power available, a magnetic field must be produced and turned off to turn the electric motor on. The generators are responsible for the generation of this reactive power at all the time for the consumer. There are cycles of magnetic fields to produce reactive power which in some cases can go up to fifty cycles per day. In short, in the time of twenty-four hours, the power plants need to operate fifty times between the cycles of pumping water, generating active power and generating reactive power. In underdeveloped and developing nations, pumped storage systems are utilized mostly to meet the peak demand for electricity, since there is a shortage of power, and electric circuits are used to provide the reactive power (Belessiotis & Delyannis, 2000; Bilgen et al., 2008).

While a storage power plant and run-of-river power plant without upward pumping of the water are producers of energy, a pumped storage plant is an energy consumer which works to cater peaks demand with no influx into the upper reservoir. Power is needed to pump the water up the mountain, possibly from a coal-based power plant. There are limitations in the pump systems and pumping of water also involves energy losses. There are quite significant losses involved in streaming the water down the mountain, in friction of pipes, and in driving the turbines. According to laws of physics, about 20 kWh or more will be lost in the form of heat from every 100 kWh of electrical production taken to power the pump from the grid;

evidently, 80 kWh can be served by the producer into the grid as electric power (see Figure 1.7).

Multiple individual efficiencies can be expressed as the overall efficiency of a hydropower plant as shown by the equation given:

$G_{overall}$= gtransformer x ggenerator x gturbine x gpumps x gothers

Using the diversion of potential energy, pumped storage plants are the way to store hue amounts of electrical energy. Other storage mediums have much lesser storage capacities compared to pumped storage plants. The key reason behind the discussion on pumped storage plants in the background of the utilization of changing sources of renewable energy like wind stays in its huge storage capacity (Tsai & Chen, 2014; Manzano et al., 2017). On the one hand, pumped storage plants allow the pumping of water to the upper basin with the assistance of additional power by the wind energy system which would allow separation of the wind power supply by the consumer. On the other hand, the number of pumped storage power plants can be increased to fulfil the demands which were being met by the conventional power stations in case of major fluctuations of the wind.

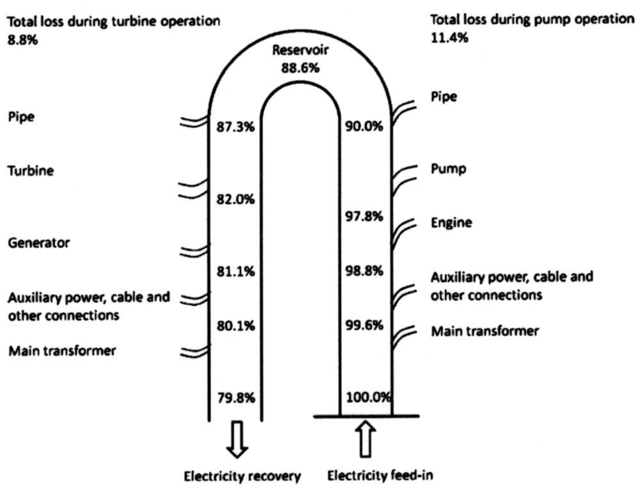

Figure 1.7. Sample energy balance of a pumped-storage power plant.

[Source: https://www.springer.com/gp/book/9783642207082.]

However, it is worth noting that the good sites for wind are usually far from the low mountain ranges in which the pumped storage plants can be constructed with viable efficiency.

1.4.4. Oceanic Power Plants

Oceanic power plants are another kind of hydropower plants as represented in Figure 1.3. There is a huge diversity of the kinds of construction shown by this category of power plants.

To begin with, the tidal power plant has a dam, like for a river power station having a dam. However, in this kind of power station, water is stored behind a dam during the high tide from the ocean side, and also during the low tide, it is stored from the opposite side. Propeller and Kaplan turbines are used in this type of power plant (Tsai & Chen, 2014; Manzano et al., 2017).

The wave power plant is a different construction kind of oceanic power plant. The potential power of waves is converted to electrical output in wave power plants through a mechanism shown in Figure 1.8. Using a pendulum door transforms the wave energy in pressure of fluid (left scheme). Another mechanism uses the wave water-pressed air that passes from a wind propeller.

Another type, the oceanic heat power plant, functions on the principle of "Ocean Thermal Energy Conversion"(OTEC). In this mechanism, the difference in temperature in deep water and surface water is used to generate electricity through a circuit. However, nowadays, there is no commercial OTEC power station in process.

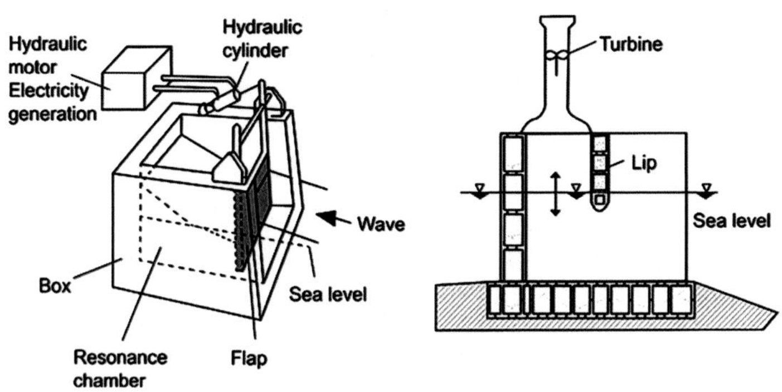

Figure 1.8. Mechanism of wave power plants (right Trivandrum/India, left Yagishiri/Japan).

[Source: https://link.springer.com/chapter/10.1007/978–3-642–20709–9_1.]

The current power plant is another kind of oceanic power plant. This power plant has the same mechanism as of wind energy power plant but underwater. The energy from flowing water is used just like wind energy to electricity by rotors. The difference in water densities arising from the temperature difference is the main driver of the current power plants using the movement of water in oceans (Manzano et al., 2017).

The osmotic power plant is yet another oceanic power plant construction. The change in salinity of the river water and ocean water is used to gain energy that generates pressure in the water. An expansion engine is sued to covert the pressure to rotating energy to drive the generator.

The electricity generation in the oceanic power plants is yet in the technical development stage except for the tidal oceanic power generation.

1.4.5. Classification on the Basis Power Production Capacity

There is a varying range of facilities based on power generation capacity from micro and small stations that individuals function for their private energy requirements or to trade power to service companies to large power plants that provide millions of customers with electricity. According to the power generation capacity, the three categories of hydropower plants are given as follows (Andre, 1978; Loots et al., 2015):

- micro hydropower;
- small hydropower; and
- large hydropower.

The definition of these classes differ from country to country; however, commonly micro-hydropower plants take capacities of no more than 100 kW, small hydropower plants take these capacities in the range of 0.1 to 15–20 MW, and large hydropower facilities are those which have capacity more than 15–20 MW (up to several GW).

Table 1.3. Worldwide Use of Hydropower in 2006.

Country	Hydro Power Generation (TWh)	Share of total generation (%)
Austria	40	59
Brazil	350	83
Canada	360	58
China	440	15
France	60	11
Germany	30	4
India	110	15

Japan	100	9
Norway	120	99
Russia	180	18
USA	320	7

Some European countries including Spain, Ireland, and Portugal agree to take 10 MW as the higher limit for installed small capacity, though, there is no international agreement on the description of small hydropower. In Sweden, power plants having more than 1.5 MW are likely to trade their electricity at lower prices, and in Italy 3 MW and France, the limit has recently been set at 12 MW. These limits are the maximum worth of installed power for which the grid has the responsibility to purchase electricity from sources of renewable energy. For small hydropower, Great Britain accepted a threshold of 20 MW. For the feed-in terrify, the capacity of hydropower is significant in Germany. For hydropower stations, the limit is different in different countries. This limit is about 15 MW for small hydropower plants in India (Manzano et al., 2017).

1.5. WORLDWIDE STATUS OF HYDROPOWER

Hydroelectric facilities produce almost 16% of the electric energy generated globally. Some countries mainly rely on this power source for their electricity production. Canada gets 58% of its electric power from water, Brazil 84%, and Norway even 99% as presented in Table 1.3 (Hotza & Costa, 2008).

On the scale of the installed power production capacities, the world's largest storage power plants are the Three Gorges Dam in China, on the border between Brazil and Paraguay, the Itaipu Dam in South America, and many other dams are represented in Table 1.4. The output of these dams in terms of electrical energy is up to twenty five times higher than in a coal-fired power plant of a single unit. The electrical demand of Paraguay and over 20% of that of Brazil is statistically met by the output from Itaipu (Jin et al., 2019).

Table 1.4. Largest Hydropower Plants Worldwide

Power plant	**Installed capacity (MW)**
Jinping II Hydropower Plant (China)	4,800
Churchill Falls hydroelectric power plant (Canada)	5,430
Robert-Bourassa hydroelectric power plant (Canada)	5,620
Nuozhadu Dam (China)	5,850
Krasnoyarsk Hydroelectric Dam (Russia)	6,000
Xiangjiaba Dam (China)	6,000
Longtan Dam (China)	6,300
Sajano-Schuschenskaja GES (Russia)	6,400
Grand Coulee Dam (USA)	6,810
Tucurui Dam (Brazil)	8,550
Guri Dam—Simón Bolívar hydroelectric power plant (Venezuela)	10,240
Belo Monte Dam (Brazil)	11,000
Xiluodu Dam (China)	12,600
Itaipu Dam (Brazil/Paraguay)	14,000
Three Gorges Dam (China)	22,500

Table 1.5. World's Largest Pumped Hydropower Station

Power plant	**Installed capacity (Turbine power) (MW)**
Tumut 3 (Australia)	1,500
Mingtan Power Plant (Taiwan)	1,600
Kazunogawa Hydroelectric Power Station (Japan)	1,600
Raccoon Mountain Pumped Storage Station (USA)	1,600
Dinorwig Power Station (Wales)	1,730
Grand Maison (France)	1,800
Tianhuangping Pumped Storage Power Plant (China)	1,840
Ludington Pumped Storage Power Plant (USA)	1,870
Okutataragi Pumped Storage Power Plant (Japan)	1,940
The Dniester Pumped Storage Plant (Ukraine)	2,270[b]
Huizhou Hydroelectric Power Plant (China)	2,400
Guangzhou Pumped Storage Power Plants I & II (China)	2,400
Robert Moses Niagara Hydroelectric Power Station (USA)	2,520
Bath County (USA)	2,730

[a]Completion 2016.

[b]Completion not yet reached.

Hydroelectric power plants of this size have significant effects on the living space of people and also on nature. According to press releases, the Three Gorges Dam project in China, resulted in the relocation of approximately one million inhabitants. Such a huge amount of water is also a possible threat for floods wall of the dams breaks, and huge areas would be flooded and many inhabitants threatened. Renewable energies are no longer environmentally neutral when applied locally and on such a scale (Abbasi et al., 2000; Akorede et al., 2012). There are several ecological compensation measures even for run-of-river power stations. Spawning waters, the creation of flood-plain forests, or the near-natural fish stairways are included in these compensation measures.

There are various large pumped storage power plants beside the storage power plants in operation globally. The most important ones are shown in Table 1.5. By worldwide standards, even the "largest" European hydroelectric power plants are considered small. Theoretically, an electricity-producing unit should fulfil the necessities of a city of about 700,000 residents by a modern hard-coal-fired power plant which has an output of 700 MW in an developed country. In Austria the storage power plants Kaprun and Malta are in this production range. However, in Germany, the largest run-of-river hydropower plant has a production of only about 20% of this level. This difference mainly arises due to the usable height difference and the available water supply. Therefore, rivers are crowded with many power plants. There are ship sluices on the Moselle and other rivers so that the ships can avoid the change in altitude of the dammed water, by these methods, in the first place. some rivers were made maneuverable. The largest pumped hydropower plants in Europe are Goldisthal in Germany and Vianden in Luxemburg, each producing more than 1,000 MW power (Grimmond, 2007).

About 4% of the electricity consumption is provided by the hydroelectric facilities in Germany. Furthermore, further development in this source of electricity has very little scope in Germany; however, hydroelectric power facilities have huge potential in other parts of the world. The potential in this system does not represent the construction of new hydroelectric facilities, but it shows mainly the expansion and modernization of the existing systems. For example, the output of at Rheinfelden, the hydroelectric power station on the upper Rhine River can be multiply to 100 MW from 26 MW by modernizing this system (Mills, 2008). A large number of small hydroelectric power stations were reactivated in the 1990s which had been turned off previously; consequently, about 5000 small hydroelectric power plants are in operation in Germany. Although, 4% of the electricity share of Germany does not come from small hydroelectric power stations, from 120 large systems which generate more than 3825 MW of electricity, constructed on larger rivers, and functioned power producing companies. Owing to their large amount of water and size, they deliver about 80% hydroelectric power of the annual generation (see Table 1.6).

The development and condition of the hydropower zone in underdeveloped nations are dissimilar from the developed ones. Economical, technological, and social constraints in developing nations are the main hurdles in the way of hydroelectric power generation. India, for example, is gifted with huge potential for hydroelectric power and is ranked fifth in the area of table hydro-potential on a worldwide level. It is estimated that at 60% plant load factor, India has about 84,000 MW (149,000 MW installed capacity) economically viable and exploitable hydro-potential. Moreover, the cumulative potential of 94,000 MW has been assessed with the identification of 56 sites for the pumped storage scheme. Additionally, 1500 sites have been identified for an aggregate potential of 94,000 MW from micro and small schemes (Fridleifsson et al., 2008; Wani & Khan, 2012). On the whole, India is gifted with a hydro-potential of almost 250,000 MW. The key potential of hydropower is from the basins of large rivers as presented in Table 1.7. Though, because of the numerous limitations confronting the sector, utilization of hydro-potential has not been up to the preferred level. Till 2010, only almost 20% of the total exploitation potential has been utilized. The limitation that has affected the growth of hydroelectric systems vary from financial (deficits in providing long term financing), technical (inadequacies in tunneling methods, difficult investigation), managerial and tariff-related issues. These schemes are also disturbed by geological issues (particularly in the region of Himalayan wherever underground tunneling is needed).

Table 1.6. Hydropower Plants in Germany in 2005

	Number of plants	**Net capacity (MW)**	**Net generation (TWh/annum)**
Over 1 MW	400	3,400	17.3
Under 1 MW	6,900	530	2.3
Total	7,300	3,930	19.6

Table 1.7. Hydropower Potential in India Excluding Small Hydropower Stations and Pumping Hydropower Stations

Basin/Rivers	**Likely installed capacity (MW)**
Brahmaputra basin	66,000
Eastern streaming rivers of southern India	15,000
Western streaming rivers of southern India	9,000
Central Indian river system	4,000
Total	149,000

Issues due to delay in acquisition of land, inaccessibility of the area, and relocation of project disturbed families.

"Policy on Hydro Power Development" was announced by the Government of India in 1998 under which incentive is provided to the establishment of hydropower in the state. In the course of hydropower establishment, this policy is considered a longed-for step towards useful exploitation of water resources. The Central Electricity Authority (CEA), during October 2001, brought about a study that ranked the upcoming operatable projects. According to this study, contingent upon their inter-se attraction, about 400 hydro systems with a cumulative installed capacity of 107,000 MW were placed in A, B & C categories. The government of India initiated the 50,000 MW hydro projects during May 2003 in which research on Pre-Feasibility Reports of 162 Projects commuting to 50,000 MW was utilized by CEA through numerous agencies. (source: NHPC, India) (Tripathi et al., 2016). Moreover, the development of the small hydro sector has also been helped by the Clean Development Mechanism (CDM) through making their financial feasibility better in India.

1.6. PROS AND CONS OF HYDROPOWER

The pros of hydropower are given below: (Singh et al., 2014; Tripathi et al., 2016)

- The absence of radioactive waste and air pollution makes hydropower a relatively clean source of electrical energy.
- It is not dependent on fossil fuels and, hence, is considered a renewable source of energy because it is obtained from natural source, i.e., water.
- It is considered a more environment-friendly energy source because of its significantly low environmental impact.

- The projected life of hydropower plants is relatively greater than other energy sources, extending over 50 years. Hydropower plants installed at the epoch of the 19th century are still working (e.g., Darjeeling power plant set up in 1897 in India).
- The efficiency of hydropower plants is comparatively higher than gas (up to 60%) and thermal energy (up to 45%).
- Load variations, instantaneous starting, stopping, etc. of hydropower stations are distinctive inherent features of hydropower plants that help in refining the consistency of the power system. Hence, to meet the peak demands, hydro stations are an excellent choice.
- No variable operational cost is involved in hydropower plants after the initial installation.
- The typical cost of generation, maintenance and operation over a lifespan is less than any other production source of energy.
- Several associated profits of flood control, navigation, irrigation, drinking water supply, pisciculture, recreation, tourism, etc., are also provided by the storage-based hydropower systems.
- Development of the unprivileged areas as a result of the location of power plants in remote areas is another benefit of the hydroelectric power plants which can help in improving medical services, telecommunication, education, road communication, etc., of the area.

There are a few disadvantages or limitations of hydropower, particularly of large hydropower plants:

- Heavily concentrated load on the earth due to large dams can lead to seismic effects.
- Due to the accessibility of head and water, the generation of hydropower is limited to hilly or foothill areas. Additional investment is required for the installation of long transmission lines which inevitably causes transmission losses.
- Difficulties are encountered in the restoration and rehabilitation of the people and activities in submerged areas.
- Dam safety may be compromised by the heavy rains in the catchment areas during the rainy season. The discharge of a huge amount of water is not feasible at the same time since the risk of floods to downstream zones is enhanced.

- The time required for the construction of hydropower plants is higher in underdeveloped countries than in other countries. This is mainly due to the challenges in hilly areas to transport heavy machinery and equipment needed to construct the plant due to lack or resources in under-developed countries.
- Turbines blades may get damaged in some locations due to silt in the water, produced from soil erosion.Frequent maintenance is required in areas where soil erosion is of significant concern, particularly in rainy seasons.
- Uncontrolled development in the catchment areas may cause serious problems for some hydropower plants in underdeveloped and developing countries. Improved employment opportunities and living conditions in this particular area nearby the dam and the power plant are the major reasons for disturbance in water flow. Uncontrolled construction of the commercial and housing sector in the catchment area, starts to cause instabilities to the movement of water into the reservoir over time, hindering the accessibility of water for power production.

REFERENCES

1. Abbasi, S. A., & Abbasi, N. (2000). The likely adverse environmental impacts of renewable energy sources. *Applied Energy*, *65*(1–4), 121–144.
2. Akorede, M. F., Hizam, H., Ab Kadir, M. Z. A., Aris, I., & Buba, S. D. (2012). Mitigating the anthropogenic global warming in the electric power industry. *Renewable and Sustainable Energy Reviews*, *16*(5), 2747–2761.
3. Alkon, M., He, X., Paris, A. R., Liao, W., Hodson, T., Wanders, N., & Wang, Y. (2019). Water security implications of coal-fired power plants financed through China's Belt and Road Initiative. *Energy Policy*, *132*, 1101–1109.
4. Andre, H. (1978). Ten years of experience at the "La Rance" tidal power plant. *Ocean Management*, *4*(2–4), 165–178.
5. Armaroli, N., & Balzani, V. (2011). Towards an electricity-powered world. *Energy & Environmental Science*, *4*(9), 3193–3222.
6. Arutyunov, V. S., & Lisichkin, G. V. (2017). Energy resources of the 21st century: problems and forecasts. Can renewable energy sources replace fossil fuels?. *Russian Chemical Reviews*, *86*(8), 777.
7. Ayres, R. U., Ayres, L. W., & Warr, B. (2003). Exergy, power and work in the US economy, 1900–1998. *Energy*, *28*(3), 219–273.
8. Banos, R., Manzano-Agugliaro, F., Montoya, F. G., Gil, C., Alcayde, A., & Gómez, J. (2011). Optimization methods applied to renewable and sustainable energy: A review. *Renewable and Sustainable Energy Reviews*, *15*(4), 1753–1766.
9. Bejan, A. (1982). Second-law analysis in heat transfer and thermal design. In *Advances in Heat Transfer* (Vol. 15, pp. 1–58).
10. Belessiotis, V., & Delyannis, E. (2000). The history of renewable energies for water desalination. *Desalination*, *128*(2), 147–159.
11. Bertsimas, D., Litvinov, E., Sun, X. A., Zhao, J., & Zheng, T. (2012). Adaptive robust optimization for the security constrained unit commitment problem. *IEEE Transactions on Power Systems*, *28*(1), 52–63.
12. Bilgen, S., Keleş, S., Kaygusuz, A., Sarı, A., & Kaygusuz, K. (2008). Global warming and renewable energy sources for sustainable development: a case study in Turkey. *Renewable and Sustainable Energy Reviews*, *12*(2), 372–396.

13. Boyd, C. E. (2012). Water quality. *Aquaculture: Farming Aquatic Animals and Plants, Second Edition.* Lucas JS and Southgate PC (eds). Blackwell Publishing Ltd, 52–83.
14. Boyd, C. E. (2020). An Overview of Hydrology and Water Supply. *Water Quality*, 41–63.
15. Brown, M. T., & Ulgiati, S. (2004). Energy quality, emergy, and transformity: HT Odum's contributions to quantifying and understanding systems. *Ecological Modelling*, *178*(1–2), 201–213.
16. Carrillo, A. M. R., & Frei, C. (2009). Water: A key resource in energy production. *Energy Policy*, *37*(11), 4303–4312.
17. Chakraborty, A. (2011). Advancements in power electronics and drives in interface with growing renewable energy resources. *Renewable and Sustainable Energy Reviews*, *15*(4), 1816–1827.
18. da Silva Soito, J. L., & Freitas, M. A. V. (2011). Amazon and the expansion of hydropower in Brazil: Vulnerability, impacts and possibilities for adaptation to global climate change. *Renewable and Sustainable Energy Reviews*, *15*(6), 3165–3177.
19. Dallinger, B., Schwabeneder, D., Lettner, G., & Auer, H. (2019). Socio-economic benefit and profitability analyses of Austrian hydro storage power plants supporting increasing renewable electricity generation in Central Europe. *Renewable and Sustainable Energy Reviews*, *107*, 482–496.
20. de Lima Andrade, A., & dos Santos, M. A. (2015). Hydroelectric plants environmental viability: Strategic environmental assessment application in Brazil. *Renewable and Sustainable Energy Reviews*, *52*, 1413–1423.
21. Dincer, I. (2002). The role of exergy in energy policy making. *Energy Policy*, *30*(2), 137–149.
22. Dincer, I., & Cengel, Y. A. (2001). Energy, entropy and exergy concepts and their roles in thermal engineering. *Entropy*, *3*(3), 116–149.
23. Ebrahimi, K., Jones, G. F., & Fleischer, A. S. (2014). A review of data center cooling technology, operating conditions and the corresponding low-grade waste heat recovery opportunities. *Renewable and Sustainable Energy Reviews*, *31*, 622–638.
24. Ellabban, O., Abu-Rub, H., & Blaabjerg, F. (2014). Renewable energy resources: Current status, future prospects and their enabling technology. *Renewable and Sustainable Energy Reviews*, *39*, 748–764.

25. Fazeli, A., Bakhtvar, F., Jahanshaloo, L., Sidik, N. A. C., & Bayat, A. E. (2016). Malaysia′ s stand on municipal solid waste conversion to energy: A review. *Renewable and Sustainable Energy Reviews*, *58*, 1007–1016.
26. Fickett, A. P., Gellings, C. W., & Lovins, A. B. (1990). Efficient use of electricity. *Scientific American*, *263*(3), 64–75.
27. Fridleifsson, I. B., Bertani, R., Huenges, E., Lund, J. W., Ragnarsson, A., & Rybach, L. (2008, January). The possible role and contribution of geothermal energy to the mitigation of climate change. In *IPCC Scoping Meeting on Renewable Energy Sources, Proceedings, Luebeck, Germany* (Vol. 20, No. 25, pp. 59–80)
28. Gagliano, A., Tina, G. M., Nocera, F., & Patania, F. (2014). Technical and economic perspective for repowering of micro hydro power plants: A case study of an early XX century power plant. *Energy Procedia*, *62*, 512–521.
29. Gartner, E. (2004). Industrially interesting approaches to "low-CO2" cements. *Cement and Concrete Research*, *34*(9), 1489–1498.
30. Georgescu-Roegen, N. (1975). Energy and economic myths. *Southern Economic Journal*, 347–381.
31. Grimmond, S. U. (2007). Urbanization and global environmental change: local effects of urban warming. *Geographical Journal*, *173*(1), 83–88.
32. Hasnain, S. M. (1998). Review on sustainable thermal energy storage technologies, Part I: heat storage materials and techniques. *Energy Conversion and Management*, *39*(11), 1127–1138.
33. Hennig, T., Wang, W., Feng, Y., Ou, X., & He, D. (2013). Review of Yunnan's hydropower development. Comparing small and large hydropower projects regarding their environmental implications and socio-economic consequences. *Renewable and Sustainable Energy Reviews*, *27*, 585–595.
34. Hotza, D., & Da Costa, J. D. (2008). Fuel cells development and hydrogen production from renewable resources in Brazil. *International Journal of Hydrogen Energy*, *33*(19), 4915–4935.
35. Huang, Y., Zhang, X., Ma, Z., Zhou, Y., Zheng, W., Zhou, J., & Sun, C. Q. (2015). Hydrogen-bond relaxation dynamics: resolving mysteries of water ice. *Coordination Chemistry Reviews*, *285*, 109–165.

36. Ibrahim, H., Ilinca, A., & Perron, J. (2008). Energy storage systems—Characteristics and comparisons. *Renewable and Sustainable Energy Reviews*, *12*(5), 1221–1250.
37. Jean-Baptiste, P., & Ducroux, R. (2003). Energy policy and climate change. *Energy policy*, *31*(2), 155–166.
38. Jin, Y., Behrens, P., Tukker, A., & Scherer, L. (2019). Water use of electricity technologies: A global meta-analysis. *Renewable and Sustainable Energy Reviews*, *115*, 109391.
39. Kadier, A., Kalil, M. S., Pudukudy, M., Hasan, H. A., Mohamed, A., & Hamid, A. A. (2018). Pico hydropower (PHP) development in Malaysia: Potential, present status, barriers and future perspectives. *Renewable and Sustainable Energy Reviews*, *81*, 2796–2805.
40. Kalogirou, S. A. (2004). Solar thermal collectors and applications. *Progress in Energy and Combustion Science*, *30*(3), 231–295.
41. Kammen, D. M., & Kirubi, C. (2008). Poverty, energy, and resource use in developing countries. *Annals of the New York Academy of Sciences*, *1136*(1), 348–357.
42. Karhinen, S., & Huuki, H. (2019). Private and social benefits of a pumped hydro energy storage with increasing amount of wind power. *Energy Economics*, *81*, 942–959.
43. Kaygusuz, K., & Sarı, A. (2003). Renewable energy potential and utilization in Turkey. *Energy Conversion and Management*, *44*(3), 459–478.
44. Khan, N., Dilshad, S., Khalid, R., Kalair, A. R., & Abas, N. (2019). Review of energy storage and transportation of energy. *Energy Storage*, *1*(3), e49.
45. Kubiak-Wójcicka, K., & Szczęch, L. (2021). Exploitation of Rivers in Poland for Electricity Production–Current Condition and Perspectives for Development. In *Management of Water Resources in Poland* 2(3), 327–355.
46. Loots, I., Van Dijk, M., Barta, B., Van Vuuren, S. J., & Bhagwan, J. N. (2015). A review of low head hydropower technologies and applications in a South African context. *Renewable and Sustainable Energy Reviews*, *50*, 1254–1268.
47. Manzano-Agugliaro, F., Taher, M., Zapata-Sierra, A., Juaidi, A., & Montoya, F. G. (2017). An overview of research and energy evolution for small hydropower in Europe. *Renewable and Sustainable Energy*

Reviews, *75*, 476–489.

48. Mills, D. (2004). Advances in solar thermal electricity technology. *Solar Energy*, *76*(1–3), 19–31.
49. Mukherjee, K. (2008). Energy use efficiency in US manufacturing: A nonparametric analysis. *Energy Economics*, *30*(1), 76–96.
50. Nababan, S., Muljadi, E., & Blaabjerg, F. (2012, June). An overview of power topologies for micro-hydro turbines. In *2012 3rd IEEE International Symposium on Power Electronics for Distributed Generation Systems (PEDG)* 4(4), 737–744).
51. Nõges, T., Nõges, P., & Cardoso, A. C. (2010). Review of published climate change adaptation and mitigation measures related with water. *Scientific and Technical Research Series EUR*, *24682*.
52. Patterson, M. G. (1996). What is energy efficiency?: Concepts, indicators and methodological issues. *Energy policy*, *24*(5), 377–390.
53. Pérez-Sánchez, M., Sánchez-Romero, F. J., Ramos, H. M., & López-Jiménez, P. A. (2017). Energy recovery in existing water networks: Towards greater sustainability. *Water*, *9*(2), 97.
54. Richter, B. D., & Thomas, G. A. (2007). Restoring environmental flows by modifying dam operations. *Ecology and Society*, *12*(1).
55. Şahin, A. D. (2004). Progress and recent trends in wind energy. *Progress in Energy and Combustion Science*, *30*(5), 501–543.
56. Saidur, R., Islam, M. R., Rahim, N. A., & Solangi, K. H. (2010). A review on global wind energy policy. *Renewable and Sustainable Energy Reviews*, *14*(7), 1744–1762.
57. Singh, R. R., Chelliah, T. R., & Agarwal, P. (2014). Power electronics in hydro electric energy systems–A review. *Renewable and Sustainable Energy Reviews*, *32*, 944–959.
58. Smith, C., Hill, A. K., & Torrente-Murciano, L. (2020). Current and future role of Haber–Bosch ammonia in a carbon-free energy landscape. *Energy & Environmental Science*, *13*(2), 331–344.
59. Storli, P. T., & Lundström, T. S. (2019). A New Technical Concept for Water Management and Possible Uses in Future Water Systems. *Water*, *11*(12), 2528.
60. Suberu, M. Y., Mustafa, M. W., Bashir, N., Muhamad, N. A., & Mokhtar, A. S. (2013). Power sector renewable energy integration for expanding access to electricity in sub-Saharan Africa. *Renewable and Sustainable Energy Reviews*, *25*, 630–642.

61. Sun, C. Q., & Sun, Y. (2016). The attribute of water. *Springer Series in Chemical Physics*, *13*, 365–368.
62. Toan, P. K., Bao, N. M., & Dieu, N. H. (2011). Energy supply, demand, and policy in Viet Nam, with future projections. *Energy Policy*, *39*(11), 6814–6826.
63. Tripathi, L., Mishra, A. K., Dubey, A. K., Tripathi, C. B., & Baredar, P. (2016). Renewable energy: An overview on its contribution in current energy scenario of India. *Renewable and Sustainable Energy Reviews*, *60*, 226–233.
64. Tripathy, S. C., Kalantar, M., & Rao, N. D. (1993). Wind turbine driven self-excited induction generator. *Energy Conversion and Management*, *34*(8), 641–648.
65. Tsai, J. S., & Chen, F. (2014). The conceptual design of a tidal power plant in Taiwan. *Journal of Marine Science and Engineering*, *2*(2), 506–533.
66. Van de Bor, D. M., Ferreira, C. I., & Kiss, A. A. (2015). Low grade waste heat recovery using heat pumps and power cycles. *Energy*, *89*, 864–873.
67. Verán-Leigh, D., & Vázquez-Rowe, I. (2019). Life cycle assessment of run-of-river hydropower plants in the Peruvian Andes: a policy support perspective. *The International Journal of Life Cycle Assessment*, *24*(8), 1376–1395.
68. Vihma, T., Screen, J., Tjernström, M., Newton, B., Zhang, X., Popova, V., ... & Prowse, T. (2016). The atmospheric role in the Arctic water cycle: A review on processes, past and future changes, and their impacts. *Journal of Geophysical Research: Biogeosciences*, *121*(3), 586–620.
69. Von Sperling, E. (2012). Hydropower in Brazil: overview of positive and negative environmental aspects. *Energy Procedia*, *18*, 110–118.
70. Wani, S. A., & Khan, D. S (2012). Indus Water Treaty-A Problem in the Utilization of Hydro-Power Resources of J & K. *IOSR Journal Of Humanities And Social Science (IOSR-JHSS) 20*, 36–42.
71. Warr, B., Ayres, R., Eisenmenger, N., Krausmann, F., & Schandl, H. (2010). Energy use and economic development: A comparative analysis of useful work supply in Austria, Japan, the United Kingdom and the US during 100 years of economic growth. *Ecological Economics*, *69*(10), 1904–1917.

72. Warr, B., Schandl, H., & Ayres, R. U. (2008). Long term trends in resource exergy consumption and useful work supplies in the UK, 1900 to 2000. *Ecological Economics*, *68*(1–2), 126–140.
73. Waters, S., & Aggidis, G. (2016). Tidal range technologies and state of the art in review. *Renewable and Sustainable Energy Reviews*, *59*, 514–529.
74. Wood, G., & Newborough, M. (2003). Dynamic energy-consumption indicators for domestic appliances: environment, behaviour and design. *Energy and Buildings*, *35*(8), 821–841.
75. Yah, N. F., Oumer, A. N., & Idris, M. S. (2017). Small scale hydro-power as a source of renewable energy in Malaysia: A review. *Renewable and Sustainable Energy Reviews*, *72*, 228–239.
76. Yaseen, Z. M., Ameen, A. M. S., Aldlemy, M. S., Ali, M., Abdulmohsin Afan, H., Zhu, S., ... & Tao, H. (2020). State-of-the Art-Powerhouse, Dam Structure, and Turbine Operation and Vibrations. *Sustainability*, *12*(4), 1676.
77. Zeng, Q., Yan, T., Wang, K., Gong, Y., Zhou, Y., Huang, Y., ... & Zou, B. (2016). Compression icing of room-temperature NaX solutions (X= F, Cl, Br, I). *Physical Chemistry Chemical Physics*, *18*(20), 14046–14054.
78. Zhang, X., Huang, Y., Ma, Z., Zhou, Y., Zheng, W., Zhou, J., & Sun, C. Q. (2014). A common supersolid skin covering both water and ice. *Physical Chemistry Chemical Physics*, *16*(42), 22987–22994.
79. Zhang, X., Zhou, Y., Gong, Y., Huang, Y., & Sun, C. (2017). Resolving H (Cl, Br, I) capabilities of transforming solution hydrogen-bond and surface-stress. *Chemical Physics Letters*, *678*, 233–240.

Chapter 2

Small Hydropower Technology

CONTENTS

2.1. INTRODUCTION

Renewable energy sources, with their clean and sustainable production methods, can help to mitigate the environmental impacts that power generation using non-renewable energy resources inherently has. Alternative energy sources include wind, hydro, biomass, geothermal, and solar all of which are already present in different countries' energy grids.

A technology utilized in the renewable form of little hydropower, according to the United Nations Industrial Development Organization (UNIDO), allows for the advancement of rural communities and the approach to energy for a slice of the community inhabiting in these certain areas, as well as contributing to social incorporation and long-lasting growth. These are some of the factors that contribute to a positive assessment of government policies (Liu et al., 2013). The deployment and maintenance costs of hydroelectric generation are high as well as some short-term limitations exist. In the longer term, it becomes appealing for a sustainable and clean generation as well as the advantage of being utilized extremely near to high consumer centers, lowering the prices of distribution, for instance.

2.2. PLANTS OF SMALL HYDROPOWER

Small hydropower plant (SHP) definitions have no international acceptance and their categorization is completely dependent on a country's level of hydroelectric expansion. **Table 2.1** below depicts the definitions and classifications used in some of the world's most prominent small hydropower generators.

There are three different types of hydropower plants (Okot, 2013):

- Impoundment: This hydroelectric system uses a dam system to store water in a large reservoir for significant generation.
- Diversion: A canal or penstock must be built to allow a section of the river to flow to the producing group in order to generate energy with diversion. The system of dam cannot be completely reliant on the system of diversion.
- Run-of-river: This system makes advantage of the river's natural flow, obviating the necessity for impoundment in some situations.

Table 2.1. Small Hydropower Plant (SHP) Classification and Definition in Some Countries (Ferreira et al., 2016)

Country/organization	**Small (kW)**	**Mini (kW)**	**Micro (kW)**
China	≤50,000	≤2000	≤100
USA	<15,000	501–2000	<500
Japan	<10,000	-	-
Sweden	101–15,000	-	-
Philippines	<15,000	51–500	-
India	-	<2000	<100
Brazil	1001–30,000	101–1000	<100
Nigeria	-	501–2000	≤500
New Zealand	<50,000	<10,000	-
France	<50,000	501–2000	<500
Canada	1001–1,500	<1000	-
United Kingdom	-	-	<1000
Norway	1000–10,000	101–1000	<100
Russia	<30,000	-	-
Turkey	<10,000	101–2000	<100
Germany	<12,000	501–2000	<500

The preference for little technology of hydropower plants is dependent on the utilization of a run-of-river system, which utilizes the kinetic power of flowing water to move turbines and requires few or no dams on the place that keeps the project of hydropower. The run-of-river system mitigates the negative impacts of large hydroelectric dams in the installation area of the plant, such as changes in river composition and temperature and farmland flooding (Kosnik, 2010).

2.2.1. Components and Characteristic

Figure 2.1 illustrates typical illustration of run-of-river small hydropower scheme.

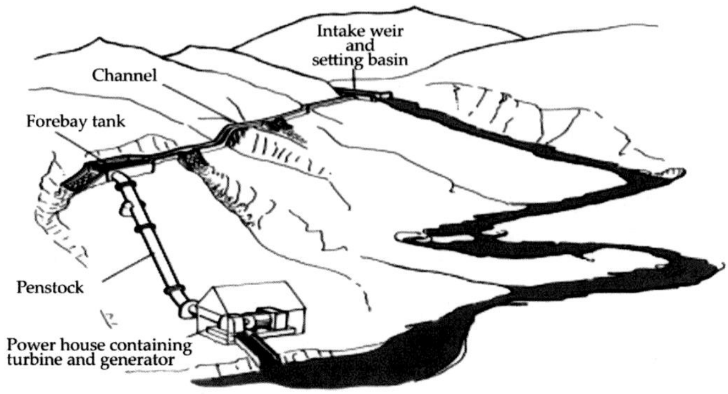

Figure 2.1. Layout of typical hydropower site.

[Source: https://www.sciencedirect.com/science/article/abs/pii/S1364032115013003.]

The weir, a small canal, the penstock, or "leat," and the settling tank (forebay) are the basic components. An intake at the weir diverts water from the main river's passage. The weir is an artificial barrier that controls the flow of water via the intake. Water is passed through a settling tank to remove particulates before entering the turbine. In the settling tank, the water is adequately decelerated to allow particulate matter to settle out. To safeguard the turbines from larger materials like man-made litter, timber, stones, and leaves that may be found in the waterway, a metal bars protective rack(trash rack) is found close to the forebay(Okot, 2013).

To comprehend the parameters that influence the advantages of an SHP, one must first comprehend the role of the key components in a hydroelectric project. The following factors are remarkable for small hydropower plants (Kim et al., 2015).

- A dam is a plant structure that elevates and maintains the engine room upstream level, artificially producing a local irregularity.
- Water intake channels, leakage channels, low-pressure adduction tunnels, pipes, external or underground powerhouses, high-pressure ducts, any surge shafts or load chambers, and tunnels are all part of the generation circuit. The designed generation circuit is to introduce water into the process of transforming mechanical energy to electrical.
- A spillway was constructed to reduce the more important design flow required to sustain the reservoir's demanded water level

while avoiding the chance of water entering the dam crest. This is the dam's protection system.

- We have a generation circuit such as:
- Penstock: The structure that connects the intake of water to the under-pressure powerhouse. External or passageway penstocks are available.
- Water intake: structure for capturing water and transporting it to the adduction tunnel or penstock.
- Powerhouse: The electrical and mechanical equipment is housed in this structure. The kind of generator and turbine determines the typical powerhouse configuration, as it does with any project of this nature.
- Equilibrium chimney: aims to avoid pressure changes caused by total or partial changes in load variations, water outflow in initial conditions, or generating unit load shedding.
- Channel and adduction tunnel: Adducting water toward the forced comportment within arrangements of a shunt is the responsibility of structures.
- Tunnel or tailrace: The channel by which the water turbinated is drained and reverted to the river is installed at the bottom of the suction duct between the powerhouse and the estuary.
- Load chamber: The change within the penstock and channel the water intake is made by this structure. It is designed to withstand critical starting conditions as well as a sudden shutdown of the generating unit.

2.2.2. Project Steps

The utilization of hydropower potential is governed by institutional, commercial, and environmental rules. Multidisciplinary practices are mixed all across the project execution process, forming the legal structure for the overall project. Figure 2.2 depicts the interdisciplinary nature of studies by illustrating the activities that are ideal for the study of an SHP and growth.

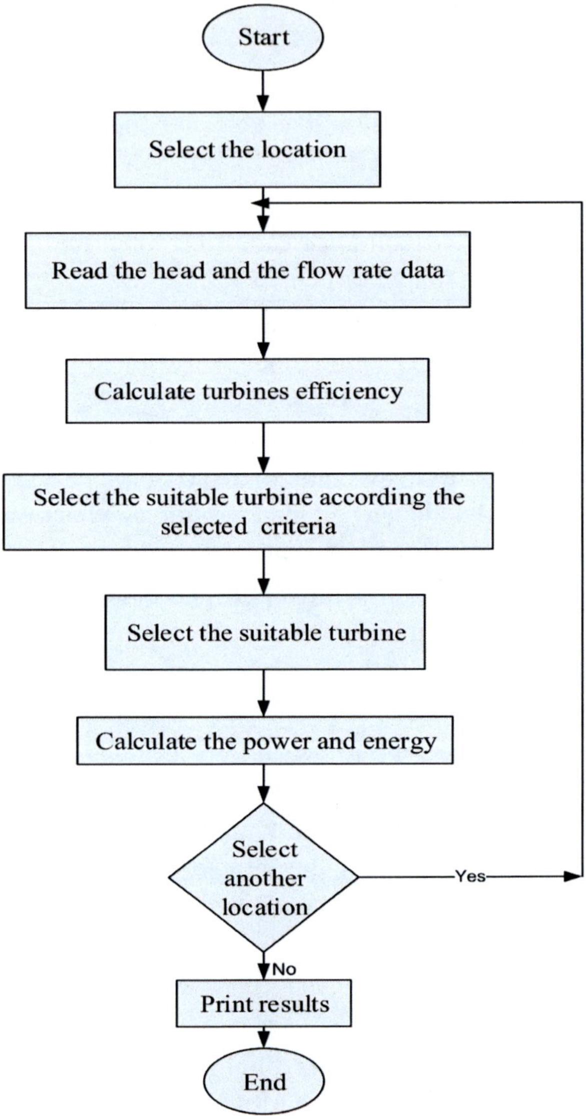

Figure 2.2. Small Hydropower Plant project steps

[Source: https://www.sciencedirect.com/science/article/pii/S2211467X19300409.]

A step cycle for implementing a project that focuses to use a hydropower project for electricity production includes phases for estimating, executing, and planning the project.

According to reference (Teixeira et al., 2004), these phases are as follows:

- Viability: The studies are more detailed here, with a focus on technical, economic, energy, and environmental viability, intending to determine the best use of the power auction. The research findings include on-site field investigations, as well as the layout of the reservoir's use and its sphere of influence, as well as the regional infrastructure projects required for its execution. Analyze the water various uses as well as the effects on the environment. We prepared an enterprise-specific environmental impact report (EIR) and environmental impact assessment (EIA) based on these research findings to obtain a preliminary license (PL) from environmental agencies.
- Hydropower potential estimation: This step entails a preliminary examination of the river basin's characteristics, particularly in terms of geological, topographic, environmental, and hydrological aspects, and to confirm the river basin's ability to generate electricity. Such analysis, which is done in the office and is based only on existing data, enables for a preliminary evaluation of the watershed's capability and price estimate, as well as prioritization of the next step.
- Hydroelectric inventory: This is defined by the modeling and development of different water resource alternatives of falling division, which are comprised of several projects that are compared to determine which one offers the best balance of deployment costs, environmental impacts, and energy benefits. This evaluation is dependent on secondary data, which is completed with field data and is directed by fundamental research in geology, cartography, hydrology, energy, multiple water uses, geotechnical engineering, and environmental science. The main features of this analysis are cost/benefit, indexes, and environmental indices, which will happen in the exploitations set. It is part of an inventory study that submits alternative uses for the research assessment of integrated environments to help with the process of licensing. Then added these exploitations to the nation's record of inventoried uses, which can be used to create the expansion plans explained above.

- Basic design: The structure of the feasibility studies is defined in depth in order to more clearly describe the project's technical characteristics, including electromechanical and civil equipment technical specifications, as well as programs of environmental and social. The basic project of environment should be produced to specify the contained recommendations in the EIA in order to get the installation license (IL) for the contracting of works.
- Executive project: it involves the creation of drawings that detail the electromechanical devices and civil works needed to complete the project and install the devices. In this stage, each necessary step for the construction of reservoirs is taken, comprising the execution of environmental plans to decrease or neutralize environmental destruction, and an operating permit should be ordered.

2.3. PROJECT COSTS

To reduce installation charges and enhance generation of power, it is critical to assess the existing conditions at a hydroelectric plant's installation site. The cost of installation varies depending on the location of the installation, the infrastructure available, and the capacity of the generator. The cost of a plant is influenced by a variety of factors, including the equipment. Even though they are smaller, small plants have higher installation costs (Okot, 2013).

According to Hosseini et al. (2005) and Forouzbakhsh et al. (2007), all construction, maintenance costs, and operation must be divided into two categories: annual costs and investments, while the study phases of an SHP. Civil structures, electrical equipment, mechanical equipment, towers of transmission, and some other costs(indirect) are all included in investment costs. . Annual costs are already essential for component and equipment maintenance, replacement, operation, and prevention (Hosseini et al., 2005; Forouzbakhsh et al., 2007).

2.3.1. Investment Costs

As mentioned below (Hosseini et al., 2005; Forouzbakhsh et al., 2007). direct costs encompass electro-mechanical devices expenses, line of power transmission expenses, and civil costs:

1. Civil expenses are calculated for the structural features of a plant's design and construction, such as the penstock, dam, tailrace channel, and forebay, as well as other features that are developed during the feasibility stage.
2. Mostly during the planning phase of an SHP, actuation, turbines, substations, control systems, protective equipment and generators, and other electrical equipment are included in the costs of electromechanical equipment. The prices of an SHP's electromechanical equipment can fluctuate depending on the plant's potential.

The electromechanical equipment cost can also be calculated using equation from Ogayar & Vidal (2009) containing a small hydropower plant's power (P), and total head (H):

$$\text{Cost} = a\,P^{\,b-1}\,H^{\,c}\ (\text{€/kW}) \qquad (1)$$

where coefficients(a,b,c) are dependent on the spatial, geographic, or field time in which they are utilized.

1. Transmission line expenses encompass lines from the point of generation to the point where electricity is delivered to the substation. These expenses are determined by the SHP's generation capacity, roadways, location, infrastructure, and current systems. However, as the length of the transmission line grows, the value rises.

   Supervision and administration (S&A) cost, inflation, Engineering and Design (E&D) cost, and cost over the construction period are all included in the indirect expenses, according to Hosseini et al. (2005).
2. E&D expenses: E&D expenses are affected by factors such as the project's size and location. These costs, as well as the apparatus and civil works, are examined as a percentage of the total building expenses. These variables differ from one region to the next. According to studies, the cost of plants with little potential can be as low as 5%, while the cost of plants with high potential can be as high as 8%.
3. S&A costs: S&A costs include the cost of management operations, the cost of land acquisition, monitoring, and inspection. This cost is comparable to E & D's, and it is similarly calculated as a percentage of total building costs. Based on the project location, the values can range from 4% to 7%.

4. During all phases of the project, the rate of inflation must be taken into account. The inflation rate for the time and the upcoming few years, as estimated by the inflation rate(average) of the prior years, must be factored into deployment costs.

2.3.2. Annual Costs

Annual costs, as well as investment expenses, should be assessed to determine a project's net benefit. Operation and Maintenance (O&M), Depreciation of equipment, and refurbishment expenditures and replacement are all included in annual costs (Hosseini et al., 2005).

1. Equipment depreciation: During the project's economic planning, the wear, service life, and variables that may impact the equipment's functioning must all be considered.
2. O&M costs: the annual costs include the money invested on specialists in an project(SHP), such as taxes, salaries, consumables, and insurance. These costs are adjusted for local inflation each year. It is based on a 5% rate of inflation to correct professional charges. These expenses account for 2% of the cumulative yearly investment.
3. Expenses of renovation and replacement: some of the most important components of electromechanical, such as turbine runners and generator windings, will have to be exchanged and replaced at some point. After 25 years, it is expected that the expenses of equipment repairs and maintenance will be equal to the overall cost of the equipment. As a result, wear costs for equipment each component used in the power generation process of an SHP should be calculated.

2.3.3. Principles

The fundamental premise of hydropower is the transformation of a significant portion of the gross head(H(m)), into Energy(electrical and mechanical). Hydraulic turbines use the water pressure to transform potential into mechanical energy, which is used to power equipment like an electric generator and others. Water's energy is in direct relationship to its flow and pressure. The SHP components are depicted in Figure 2.2.

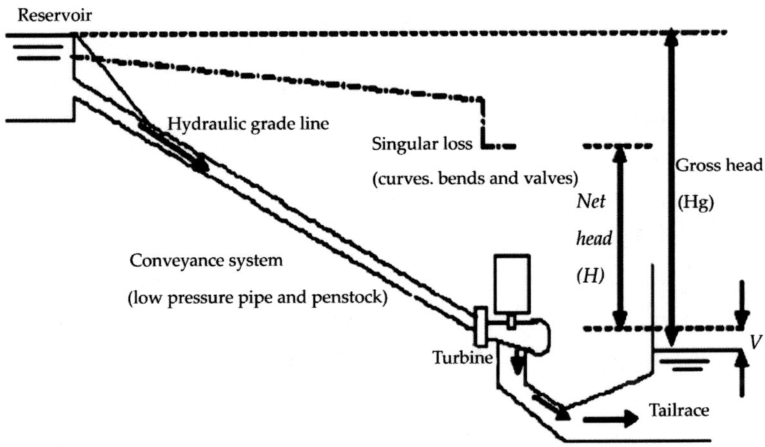

Figure 2.3. Small hydropower Components.

[Source: https://ideas.repec.org/a/eee/rensus/v11y2007i9p2152–2165.html.]

The hydraulic power(P0 in kW) and the related energy(Eo in kWh) in a time interval (Δt(h)) are generally equal to:

$$P_0 = \rho g Q H \qquad (2)$$

$$E_0 = \rho g Q H \Delta t \qquad (3)$$

where g and ρ stand for gravity acceleration (ms2) and water density (kg m3), respectively. P0 is smaller than P in terms of final power supplied to the network. Any hydroelectric plant's electricity output is determined by

$$P = \eta P_0 \qquad (4)$$

where η is the turbo-hydraulic generator's efficiency.

First, the volume and head of local plant installation and river water are measured to determine the type of hydraulic turbine for a project. Moreover, for a comprehensive and objective examination, the cost and efficiency of each available kind of turbine must be considered. For the SHP (Kim et al., 2015), there are two main types of turbines:

1. Reaction turbines—Kaplan, propeller, and Francis.
2. Impulse turbine—Turgo, Pelton, and cross flow.

2.4. TURBINE SELECTION

There are two types of hydro turbines: reaction turbines and impulse turbines. The distinction is in the manner in which energy is generated from inputs (Okot, 2013).

2.4.1. Reaction Turbine

The mutual activity of water pressure and flow generates energy in reaction turbines. They work while the rotor is immersed in water and is constrained to pressure casing. "The blades of a runner are so shaped that pressure variations along them exert lift forces, similar to that on aircraft wings, that cause the runner rotation," according to Okot (2013). Reaction turbines, unlike impulse turbines, can be used in areas with a short height drop and a larger water flow. Turbines such as Kinetic, Francis, and propeller (Okot, 2013) are examples.

2.4.1.1. Propeller Turbine

According to Okot (2013), a propeller turbine has a three to the six-bladed axial flow channel, based on the water head designed. Upon reaching the turbine hall, the water must be swirled for efficiency. Propeller turbines are best suited to locations with little water flow. Propeller turbines include Straflo, Bulb, and Kaplan

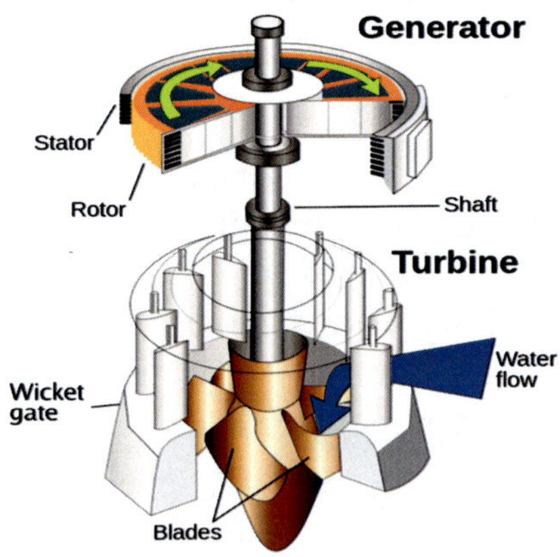

Figure 2.6. Propeller turbine schematic.

[Source: https://econpapers.repec.org/article/eeerensus/v_3a26_3ay_3a2013_3ai_3ac_3ap_3a515–520.htm.]

"For introducing inlet swirl, incorporate fixed guide vanes installed upstream of the runner and for the runner a snail shell housing, where the water flows tangentially and is pushed to spiral into the runner," according to Okot (2013). The runner blades of Kaplan turbines are set. Fixing the blades of the turbine and guiding vanes can promote efficiency greatly in a broad array of flows, but it is costly and thus only feasible in bigger systems. Propeller turbines with unregulated waterfall and flow are widely utilized when there is a possibility for little plants including waterfall and flow are reasonably steady. A typical propeller turbine is depicted in Figure 2.5.

2.4.1.2. Francis Turbine

The Francis turbine, one of the most traditional hydraulic turbine designs, has an efficiency curve that can perform in a variety of flow and height circumstances, as well as fixed runner blades and adjustable guide vanes (Johnson et al., 2015).

Okot (2013) reports that this turbine (Figure 2.6) "usually features a mixed axial/radial or radial flow runner that is most typically installed in a spiral casing with internally adjustable guiding vanes." The runner spins because the water inflow is radially through it and emerges axially. The draught tube and wicket gates, including the runner, are important parts."

When the project level is average, the turbines of Francis can have a 90% efficiency, but they can be ineffective when the measured flow at the site differs significantly from the design flow. This turbine can be mounted on a penstock or an open trough (Okot, 2013).

2.4.1.3. Kinetic turbine

Kinetic turbines use the Water kinetic energy to generate electricity utilizing the natural water flow. As a result, the systems of kinetic do not rely on artificial channels or deviations; instead, they rely on the river's natural course. They can, however, be used in such conduits (Okot, 2013). A kinetic turbine is depicted in Figure 2.7.

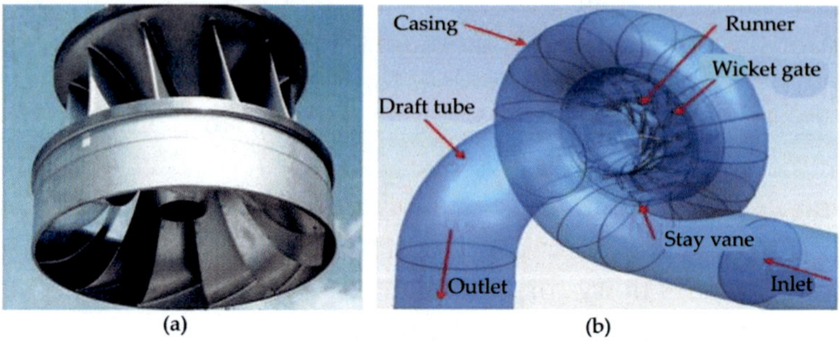

Figure 2.7. (a) Francis runner. (b) Francis schematic.

[Source:https://www.researchgate.net/publication/318702637_Prospects_of_Smal_Hydropower_Technology.]

Figure 2.8. Hydrokinetic model.

[Source: https://www.renewableenergyworld.com/baseload/ocean-tidal-stream-power-identifying-how-marine-and-hydrokinetic-devices-affect-aquatic-environments/.]

2.4.2. Impulse Turbine

The impulse turbine runner is driven by the water kinetic energy and releases to atmospheric pressure. The impulse turbine runner is operated within the air and is propelled by water jets. "The water that flows into the water tail after hitting the buckets has very little energy going to remain," according to Okot (2013). As a result, the turbine has a light housing that shields the environment from splashing of water. This turbine works well in systems with a lot of water falling but little flow. In power plants, three different kinds of impulse turbines are commonly used: turgo, Pelton, and cross flow.

2.4.2.1. Pelton

The operating head of the Pelton turbine (Figure 2.3a) is very high. Because of the high operating head, the flow rate is often low, ranging from 0.2 to 0.4 cfs. The turbine demands a high-pressure flow via the inlet, which necessitates a suitable penstock design. To concentrate the fluid flow on a runner, the Pelton uses a nozzle in the spear jet. In even tiny hydro applications, the spear buckets and jet are devised to cause limited damage resulting in a 90% potential efficiency. Up to 6 spear jets (that can be seen in Figure 2.3b) can be found in a Pelton turbine, which efficiently boosts the velocity of flow to the turbine, producing increased efficiency and power generation(Cada et al., 2012).

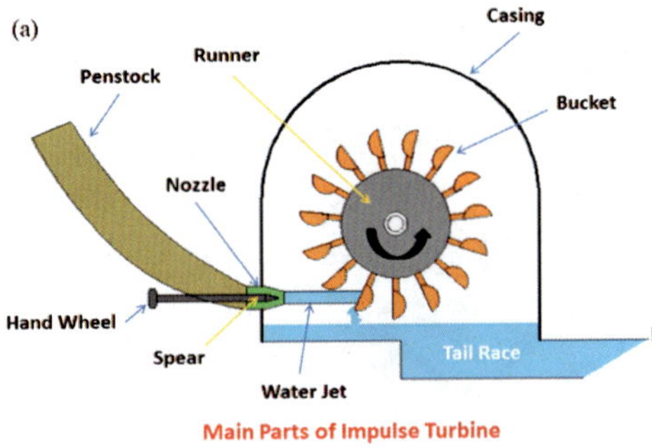

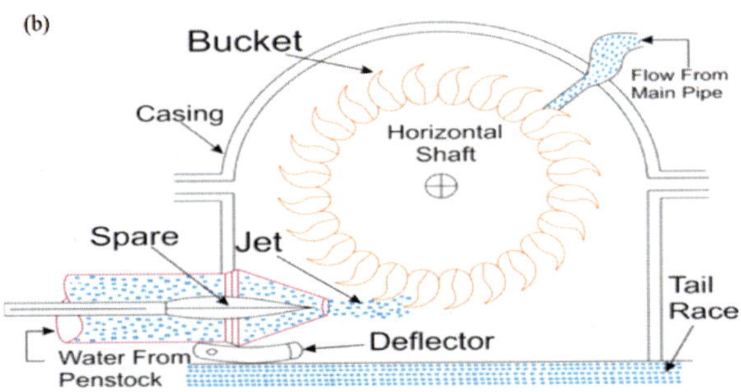

Figure 2.4. (a) Schematic of Pelton turbine. (b) Pelton turbine cross section.

[Source: https://www.mechanicaltutorial.com/pelton-turbine-or-pelton-wheel-turbine.]

2.4.2.2. Cross-Flow

The cross-flow turbine gets its name from the way water flows over the runner (Figure 2.4). This is because many cross-flow structures feature two or even more input guiding vanes at the entrance. This impulsive turbine class can handle a wide range of flow rates while maintaining great efficiency. Flow could be directed towards either a runner piece under the full runner or lower inflow when large flows are required, by changing the function of the input guide vanes to perfect match the conditions of flow. The cross-flow is capable of maintaining a steady efficiency, as seen by the efficiency curve (Cada et al., 2012)..

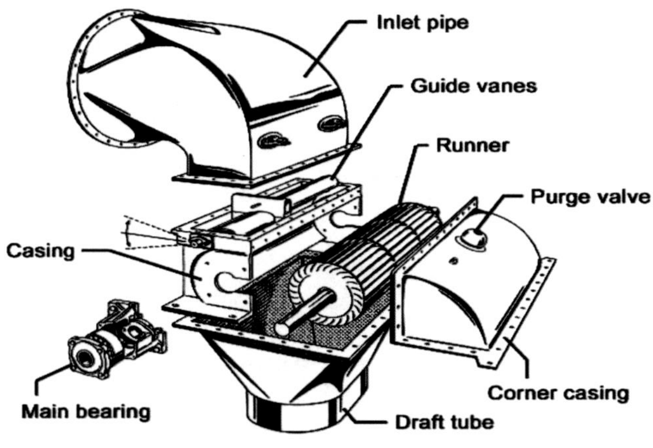

Figure 2.5. Schematic of cross-flow.

[Source:https://www.researchgate.net/publication/318702637_Prospects_of_Smal_Hydropower_Technology.]

2.4.2.3. Turgo

The Turgo and Pelton turbines are identical, but the buckets are shaped differently, as well as the jet hits the runner plane at an angle. The turgo turbine differs from the Pelton turbine in a few ways that make it a better fit for specific purposes. Because of its faster working speed, this turbine offers high overall performance and little maintenance, enabling a direct linkage between the generator and turbine. Due to the rate of flow going into this is not constrained as an inlet jet of discharged, the turgo turbine could have a less diameter than the Pelton, enhancing power generation (Okot, 2013).

2.4.3. Turbine Selection

In small plants of hydroelectric, establishing the turbine type that will be capable of function as well as the data design, can be carried out in the turbine standard size, however, there is an efficiency difference. The chart below (Figure 2.8) depicts seven primary kind of turbines and the suggested flow and range of head for each (Johnson et al., 2015).

Figure 2.8 is a graph that may be used in the introductory studies stage to determine which class of turbine to employ based on the flow of water at the installation location of a plant and project height as well as to assess their hydroelectric potential. If the local is placed at a level of 100ft with a 100 cfs flow, the Francis, Kaplan, and cross-flow turbines are equipped with as possibilities, every with its own set of benefits and drawbacks.

Manufacturers of hydraulic turbines provide an efficiency curve. This graph depicts the link between waterfall and flow, as well as the efficiency with which these two variables are examined. It is possible to examine any type of turbine and its behavior in various project scenarios. A turbine with a flatter efficiency curve may often run at a wide range of flow and head. Slightly narrower and steeper curves indicate a turbine intended for smaller operating ranges. The turbine efficiency graph is shown in Figure 2.9.

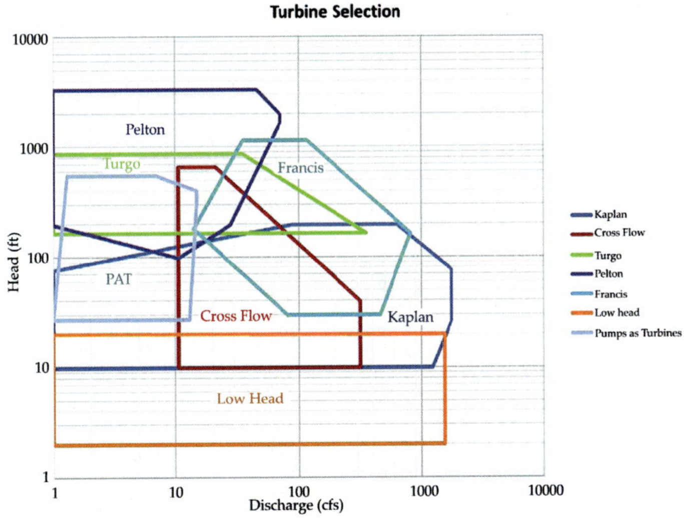

Figure 2.9. **Turbine selection chart.**

[Source: https://www.intechopen.com/books/renewable-hydropower-technologies/prospects-of-small-hydropower-technology.]

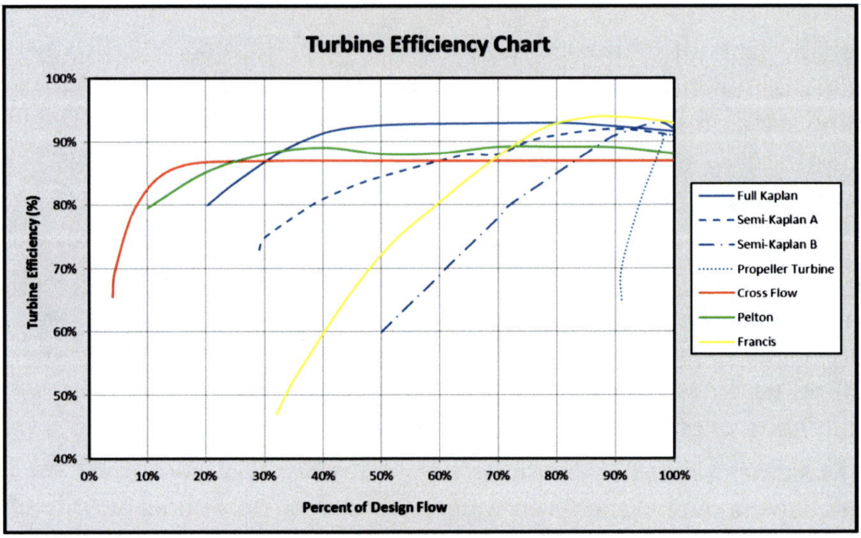

Figure 2.10. Turbine efficiency chart.

[Source: https://www.intechopen.com/books/renewable-hydropower-technologies/prospects-of-small-hydropower-technology.]

REFERENCES

1. Acar, C., & Dincer, I. (2014). Comparative Assessment of hydrogen production methods from renewable and non-renewable sources. *International journal of Hydrogen Energy*, *39*(1), 1–12.
2. Anagnostopoulos, J. S., & Papantonis, D. E. (2007). Optimal sizing of a run-of-river small hydropower plant. *Energy Conversion and Management*, *48*(10), 2663–2670.
3. Balat, H. (2007). A renewable perspective for sustainable energy development in Turkey: The case of small hydropower plants. *Renewable and Sustainable Energy Reviews*, *11*(9), 2152–2165.
4. Barbarelli, S., Amelio, M., Florio, G., & Scornaienchi, N. M. (2017). Procedure selecting pumps running as turbines in micro hydro plants. *Energy Procedia*, *126*, 549–556.
5. Benson, L., Glass, L., Jones, T. G., Ravaoarinorotsihoarana, L., & Rakotomahazo, C. (2017). Mangrove carbon stocks and ecosystem cover dynamics in southwest Madagascar and the implications for local management. *Forests*, *8*(6), 190.
6. Bertrand, K. S. E., Hamandjoda, O., Nganhou, J., & Wegang, L. (2017). Technical and economic feasibility studies of a micro hydropower plant in Cameroon for a sustainable development. *Journal of Power and Energy Engineering*, *5*(09), 64.
7. Cada, G. F., Copping, A. E., & Roberts, J. (2012). Ocean/tidal/stream power: identifying how marine and hydrokinetic devices affect aquatic environments. *Hydroworld. com (Accessed June 10, 2012).*
8. Carravetta, A., Del Giudice, G., Fecarotta, O., & Ramos, H. M. (2012). Energy production in water distribution networks: A PAT design strategy. *Water Resources Management*, *26*(13), 3947–3959.
9. Carunaiselvane, C., & Chelliah, T. R. (2017). Present trends and future prospects of asynchronous machines in renewable energy systems. *Renewable and Sustainable Energy Reviews*, *74*, 1028–1041.
10. Dyson, M., Bergkamp, G., & Scanlon, J. (2003). Flow: the essentials of environmental flows. *IUCN, Gland, Switzerland and Cambridge, UK*, 20–87.
11. Ferreira, J. H. I., Camacho, J. R., Malagoli, J. A., & Júnior, S. C. G. (2016). Assessment of the potential of small hydropower development in Brazil. *Renewable and Sustainable Energy Reviews*, *56*, 380–387.

12. Forouzbakhsh, F., Hosseini, S. M. H., & Vakilian, M. (2007). An approach to the investment analysis of small and medium hydro-power plants. *Energy Policy*, *35*(2), 1013–1024.
13. Gatte, M. T., & Kadhim, R. A. (2012). Hydro power. *Energy Conservation*, *9*(51000), 95–124.
14. Hartshorne, H., Backhouse, C. J., & Lee, W. E. (2004). Ferrofluid-based microchip pump and valve. *Sensors and Actuators B: Chemical*, *99*(2–3), 592–600.
15. Hosseini, S. M. H., Forouzbakhsh, F., & Rahimpoor, M. (2005). Determination of the optimal installation capacity of small hydro-power plants through the use of technical, economic and reliability indices. *Energy Policy*, *33*(15), 1948–1956.
16. Johnson, K., Hart, A., George, L., Young, N., & Applegate, M. (2015). Small Hydropower Handbook. *Colorado Energy Office*, *1580*, 1–50.
17. Kim, B. K., Choi, S. S., Wang, Y. P., Kim, E. S., & Rho, D. S. (2015). A study on the control method of customer voltage variation in distribution system with PV systems. *Journal of Electrical Engineering and Technology*, *10*(3), 838–846.
18. Kirkby, C. A., Giudice-Granados, R., Day, B., Turner, K., Velarde-Andrade, L. M., Dueñas-Dueñas, A., ... & Douglas, W. Y. (2010). The market triumph of ecotourism: an economic investigation of the private and social benefits of competing land uses in the Peruvian Amazon. *PloS one*, *5*(9), e13015.
19. Kong, Y., Wang, J., Kong, Z., Song, F., Liu, Z., & Wei, C. (2015). Small hydropower in China: The survey and sustainable future. *Renewable and Sustainable Energy Reviews*, *48*, 425–433.
20. Kosnik, L. (2010). The potential for small scale hydropower development in the US. *Energy Policy*, *38*(10), 5512–5519.
21. Liu, H., Esser, L. J., & Whiting, K. (2013). Realizing Rio principles through sustainable energy solutions: Application of small hydropower (SHP) in China and other developing countries. *International Journal of Technology Management & Sustainable Development*, *12*(3), 281–300.
22. Liu, X., Luo, Y., Karney, B. W., & Wang, W. (2015). A selected literature review of efficiency improvements in hydraulic turbines. *Renewable and Sustainable Energy Reviews*, *51*, 18–28.

23. Mohamed, A. R., & Lee, K. T. (2006). Energy for sustainable development in Malaysia: Energy policy and alternative energy. *Energy policy*, *34*(15), 2388–2397.
24. Mousavi, N., Kothapalli, G., Habibi, D., Khiadani, M., & Das, C. K. (2019). An improved mathematical model for a pumped hydro storage system considering electrical, mechanical, and hydraulic losses. *Applied Energy*, *247*, 228–236.
25. Nilsson, O., & Sjelvgren, D. (1997). Hydro unit start-up costs and their impact on the short term scheduling strategies of Swedish power producers. *IEEE Transactions on Power Systems*, *12*(1), 38–44.
26. Ogayar, B., & Vidal, P. G. (2009). Cost determination of the electro-mechanical equipment of a small hydro-power plant. *Renewable Energy*, *34*(1), 6–13.
27. Oh, T. H., Pang, S. Y., & Chua, S. C. (2010). Energy policy and alternative energy in Malaysia: issues and challenges for sustainable growth. *Renewable and Sustainable Energy Reviews*, *14*(4), 1241–1252.
28. Okot, D. K. (2013). Review of small hydropower technology. *Renewable and Sustainable Energy Reviews*, *26*, 515–520.
29. Parisi, M. L., Douziech, M., Tosti, L., Pérez-López, P., Mendecka, B., Ulgiati, S., ... & Blanc, I. (2020). Definition of LCA Guidelines in the Geothermal Sector to Enhance Result Comparability. *Energies*, *13*(14), 3534.
30. Prajapati, P. V. M., Patel, P. R. H., & Thakkar, P. K. H. (2015). Design, Modeling & Analysis of Pelton Wheel Turbine Blade. *vol*, *3*, 159–163.
31. Ratcheva, V. (2009). Integrating diverse knowledge through boundary spanning processes–The case of multidisciplinary project teams. *International Journal of Project Management*, *27*(3), 206–215.
32. Reich, S. M., & Reich, J. A. (2006). Cultural competence in interdisciplinary collaborations: A method for respecting diversity in research partnerships. *American Journal of Community Psychology*, *38*(1), 51–62.
33. Reid, W. V., Laird, S. A., Meyer, C. A., Gámez, R., Sittenfeld, A., Janzen, D., ... & Juma, C. (1996). Biodiversity prospecting. *Medicinal Resources of the Tropical Forest: Biodiversity and Its Importance to Human Health.* Columbia Univ. Press, New York, 142–173.

34. Schønheyder, J. F., & Nordby, K. (2018). The use and evolution of design methods in professional design practice. *Design Studies*, *58*, 36–62.
35. Seidel, V. P., & Fixson, S. K. (2013). Adopting design thinking in novice multidisciplinary teams: The application and limits of design methods and reflexive practices. *Journal of Product Innovation Management*, *30*, 19–33.
36. Shenhar, A. J., Sauser, B., Sage, A. P., & Rouse, W. B. (2009). Systems engineering management: The multidisciplinary discipline. *Handbook of Systems Engineering and Management*, *2*, 117–154.
37. Singer, S. J., Hayes, J. E., Gray, G. C., & Kiang, M. V. (2015). Making time for learning-oriented leadership in multidisciplinary hospital management groups. *Health Care Management Review*, *40*(4), 300–312.
38. Sinha, R. K., & Kakodkar, A. (2006). Design and development of the AHWR—the Indian thorium fuelled innovative nuclear reactor. *Nuclear Engineering and Design*, *236*(7–8), 683–700.
39. Teixeira, S. L. M., Bastos, F. I., Telles, P. R., Hacker, M. A., Brigido, L. F., Oliveira, C. A. D. F., ... & Morgado, M. G. (2004). HIV-1 infection among injection and ex-injection drug users from Rio de Janeiro, Brazil: prevalence, estimated incidence and genetic diversity. *Journal of Clinical Virology*, *31*(3), 221–226.
40. Tewari, V. P. (2016). Forest inventory, assessment, and monitoring, and long-term forest observational studies, with special reference to India. *Forest Science and Technology*, *12*(1), 24–32.
41. Tobi, H., & Kampen, J. K. (2018). Research design: the methodology for interdisciplinary research framework. *Quality & Quantity*, *52*(3), 1209–1225.
42. Vakis, A. I., & Anagnostopoulos, J. S. (2016). Mechanical design and modeling of a single-piston pump for the novel power take-off system of a wave energy converter. *Renewable Energy*, *96*, 531–547.
43. Walker, R., & Simmons, C. (2018). Endangered Amazon: an indigenous tribe fights back against hydropower development in the Tapajós Valley. *Environment: Science and Policy for Sustainable Development*, *60*(2), 4–15.

44. Wang, J., Chen, W., Sun, J., Liu, C., Yin, Q., Zhang, L., ... & Jiang, X. (2014). A microfluidic tubing method and its application for controlled synthesis of polymeric nanoparticles. *Lab on a Chip*, *14*(10), 1673–1677.
45. Xia, G. D., Liu, X. F., Zhai, Y. L., & Cui, Z. Z. (2014). Single-phase and two-phase flows through helical rectangular channels in single screw expander prototype. *Journal of Hydrodynamics, Ser. B*, *26*(1), 114–121.
46. Yildiz, V., & Vrugt, J. A. (2019). A toolbox for the optimal design of run-of-river hydropower plants. *Environmental Modelling & Software*, *111*, 134–152.
47. Amelio, M., & Barbarelli, S. (2004, January). A one-dimensional numerical model for calculating the efficiency of pumps as turbines for implementation in micro-hydro power plants. In *Engineering Systems Design and Analysis* (Vol. 41731, pp. 65–72).

Chapter 3

Terminologies and Legal Requirements of Hydroelectric Power Systems

CONTENTS

3.1. INTRODUCTION

A brief definition of widely used terminology relevant to power generation is needed to improve understanding and knowledge of the operation and construction of hydropower plants (Hanna et al., 201; Kuriqi et al., 2021). The following is a glossary of words used in the hydropower industry. The below terms are widely used in the context of power generation.

3.1.1. Capacity of Construction

The maximum power production capability that a river power plant can reach. The flow rate of construction at the drop of construction height determines this power (Assis et al., 2019).

3.1.2. Maximum Capacity

It's the exact capacity as the building capacity for river power plants. At the maximum adjustable capacity, all pumped storage and stored power plants has the maximum capacity at the maximum fall height (Truffer et al., 2003).

3.1.3. Rated Power

It is the machine's maximum continuous power, as agreed in the contract between the recipient and operator. It is also described as a machine's or plant's power rating, and it is typically at this loading that the machine performed better (Jiang et al., 2016).

3.1.4. Supply at Peak Load

Whenever there exists a notable electricity requirement, the power plants hydroelectric, mostly the power plants of storage, are being used to bridge the difference among supply and demand of traditional power plants. Running the machine down and up accomplishes this. The hydropower plant has the benefit of being able to ramp up and down easily. As a result, it can be used when the grid's load shifts suddenly and quickly (Karakosta et al., 2010).

3.1.5. Head and Fall Height

There is disparity in height within water level on the side of upstream (in the river power stations case or on the reservoir (upper) in the storage power stations case) and the water level on the downstream (elevation of river) by utilizing Propeller, Kaplan, and Francis turbines (Bejarano et al., 2019).

This is the disparity in height among the upstream water level and the inlet water to the turbine when utilizing Pelton turbines (nozzle).

3.1.6. Operation of Phase Shift

To drive motors or operate capacitance-based electronic devices, the consumer requires reactive power in addition to active power. During operations, long cable lines or overhead lines often need reactive power (Guittet et al., 2016). The generator is attached to the power grid to deliver only reactive power. The generator provides reactive power back to the grid by being over or under-exited. This less active power would be obtained from the grid, as well as from other power station, to offset the mechanical losses of the water empty turbine and turning a generator. Within the phase shift operation, the turbine is not powered by water.

3.1.7. Working Regularly

Regular process denotes that the storage of pumped or power plants storage is operated at the capacity-frequency-load regulator's factor. When there is a requirement for peak load availability, the pumped storage plant machines run at various capacities all of the time (Ardizzon et al., 2014). River power stations have a lower capacity shift than thermal power plants.

3.1.8. Regular Year

A regular year is an average statistical year calculated over a larger period, such as 20 to 30 years, in terms of available water (Jacobson & Delucchi, 2011).

3.1.9. Capacity Reserved

It is a supply form that is available immediately and is utilized for load changing or disruptions in different power grids or plants (Armaroli & Balzani, 2011).

3.1.10. Firm Power, Firm capacity, and Secure Power

Firm power (also known as secure power) is the operating capacity of river power plants that are guaranteed to be available for a certain number of days, such as 330 days in Germany. The number 330 was picked at random. The power is governed through the ability to separate capacity under power plants with a small-time reservoir, such as the power plant of a river. The firm's control available time is often decided upon by the operator and the receiver, for example, 5 hours a day (Patzek & Croft, 2010).

Firm power is described differently in storage power plants. The power that the power plant can generate at fall height with a 10% content of a reservoir of water volume. Firm power is produced at fall height by a 50% content of reservoir of water volume in pumped storage power plants.

3.1.11. Turbine or Generator Operation

The turbine, also known as the generator, is used to power the electric engine. The unit functions as a generator, supplying electricity to the grid.

3.2. LOAD AREAS

The requirement for electricity varies during the day over a time-dependent difference in the usage of electric devices in offices/homes and for the industrial device.

When the majority of people get ready for work in the morning, demand rises from around 6 to 8 a.m., putting a relatively heavy burden on the plants (17). Requirement, also the load on the power plants, prevails nearly constant until lunchtime (noon). Added load peak occurs early in the evening or late afternoon, as people return home and switch on their lamps, laptops, or televisions. The load gradually decreases before nightfall. As depicted in Figure 3.1, the highest load occurs in the evening or during the day at noon, while the minimum occurs at night (around 2 and 4 a.m.) (Aryandoust & Lilliestam, 2017).

It should be recorded that the diagram just depicts the modifications that a hydropower plant would deal with. With variations in climatic conditions and location, demand and variations change dramatically. For instance, within a hot land such As India, interest, and demand would also be more in the summer season due to cooling demands, whereas in Germany, interest, and demand would be more in the winter season due to demands for heating, since previously stated.

As a result, the load fluctuates not only during the day but also from one season to another season. In Germany, the energy requirement is large in the winter by the use of additional electricity and light for heating devices, for example, but a small amount of electricity is required in the summer, and thus the constant load is smaller than in the winter. In India, the pattern is in the reverse direction because the Indian weather condition is too hot(Armaroli & Balzani, 2011; Tang et al., 2018). Regardless of how much the load changes across a time, there is still a minimum load. This is referred to as baseload. on the energy that is required all day and not falls under that amount is known as constant load or baseload. The term "middle load" refers to the level of a load that is required by electricity users over numerous hours per daytime, typically of early morning to late evening. Unlike the base constant load, the intermediate load does not have a fixed level and may fluctuate with the increase in electricity consumption in the morning to the drop in the evening. Finally, the top load is the "middle load top," which can be graphically distinguished from the average load and is attained at specific times during the day.

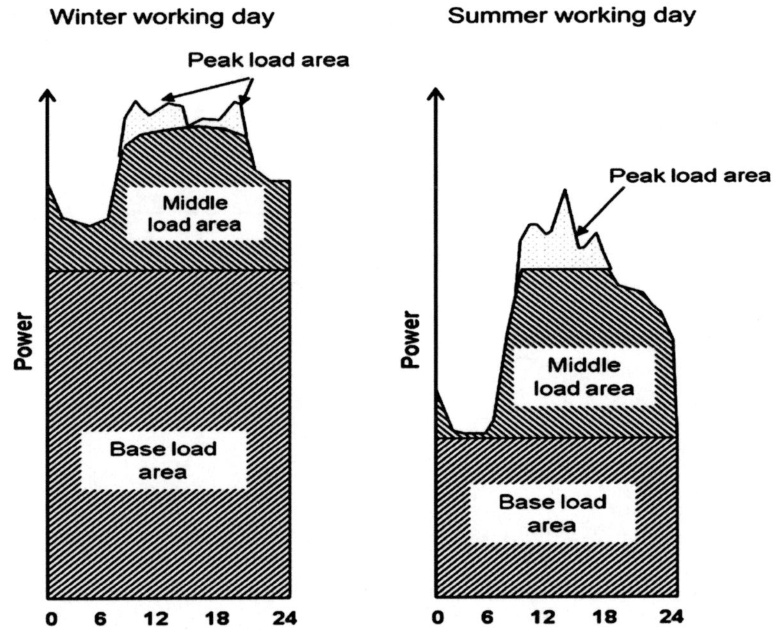

Figure 3.1. Peak, middle, and base load on a chilly country's summer and winter day. [https://link.springer.com/chapter/10.1007/978–3-642–20709–9_2.]

The hydropower plant usually has 4 distinct roles in terms of power supply. As previously mentioned, the very first responsibility is to generate power to meet the baseload. Power plants of River water are mostly responsible for this. The second task is to generate maximum load, that is mostly handled by power plants of storage. The third job of a hydropower stations in every country is to provide reactive power when it is required owing to inductive demand, which is accomplished by specifically equipped pumped storage power plants.

Finally, to satisfy seasonal fluctuations in power consumption and to offset seasonal restrictions in water supply, water must be stored (Armaroli & Balzani, 2011; Tang et al., 2018).

3.3. REACTIVE AND ACTIVE POWER

As a hydropower plant operates in the phase shifting mode, the generator sends high or low power(reactive) toward the grid (Laghari et al., 2013; Samanta et al., 2017).

As a result, the concepts of active and reactive power will be discussed further down.

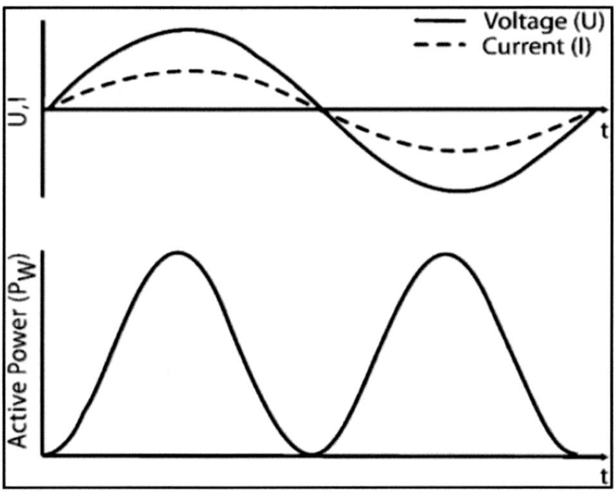

Figure 3.2. Active power with current and voltag overlapping.

[Source:https://www.researchgate.net/publication/298571181_Introduction_to_hydro_energy_systems_Basics_technology_and_operation.]

In alternating current, the time-based variability of voltage U(t) and current I(t) is sinusoidally depicted in Figure 3.2. Their time relationships are given below:

$$I(t) = I_0 \cdot \sin(\omega \cdot t - \varphi) \qquad (1)$$

and

$$U(t) = U_0 \cdot \sin(\omega \cdot t) \quad (2)$$

In the above equations: I_0 is the peak current, U_0 is the peak voltage, ω is the angular frequency (2πf), and φ represents the electric current's angular phase shift from voltage. In active power case, u is zero, as previously discuss.

The active power are:

$$P_W = \frac{1}{2} \cdot U_0 \cdot I_0 \cdot \cos\varphi \quad (3)$$

Active power is used only slightly in magnetic and electric energy storage, such as a coil with inductance L or condenser with capacity C. An alternating power is moving if it is operated with a voltage(sinusoidal alternating) of an angular frequency of x, amplitude U0, and a current of amplitude I_0.

This form of alternating power is referred to as reactive power as shown in Figure 3.3. In real, some energy is lost due to unloading and loading cycles, which result in energy transmission and hence increased resistance(Ohmic) losses in the grid. As a result, reactive loads must be prevented to the greatest extent possible, or, to put it another way, reactive power must be compensated (Armaroli & Balzani, 2011; Tang et al., 2018). The reactive power PR at sinusoidal AC voltage and alternating current is given by:

$$P_R = \frac{1}{2} \cdot U_0 \cdot I_0 \cdot \sin\varphi \quad (4)$$

with the electric power's angular phase shift u in opposition to the voltage

Electricity any end utilization necessitates the creation of an electromagnetic field; for example, a drill machine requires both reactive and active power. The portion of total power that is used as mechanical power is known as active power. The sinus curves overlap at active capacity because voltage and electric power run in the same phase.

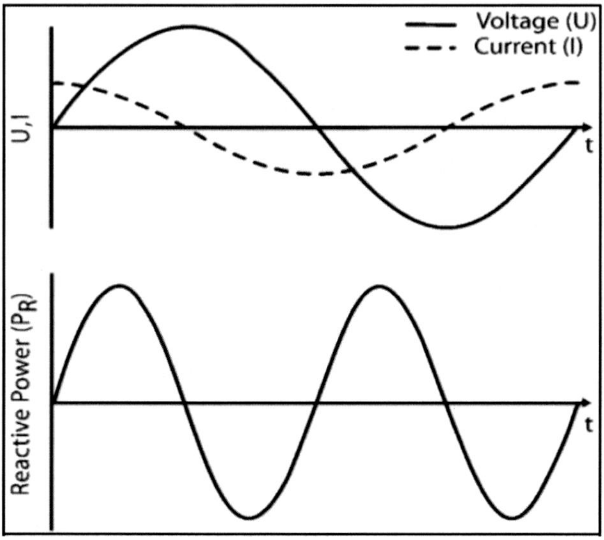

Figure 3.3. Reactive power with current and voltage not overlapping.

[Source: https://www.springer.com/gp/book/9783642207082.]

Those devices that need an electromagnetic field for functioning, on the other hand, consume reactive power. The energy within the magnetic field that rotates inside the motor is known as reactive energy. The quantity of reactive power needed is determined by the magnetic field strength required by the motor. Because a motor's shaft only can move as much as an electromagnetic field is by simply placing up and down, that energy is swayed in between electrical consumer and power plant and the with the need of reactive power. The reactive energy is used shortly by the electrical consumer before being sent backward into the power plant via the grid (Ibe & Onyema, 2013).

3.4. LEGAL REQUIREMENTS

Legal considerations are just as significant as economic and physical ones when establishing a hydropower plant. As a result, the following sections describe the various laws.

3.4.1. Water Deviation Permission

Water is necessary for both human and natural existence. For generations, having lakes, rivers or even brooks on one's property has been essential.

However, this has frequently been a source of contention. Farmers that do not have access to flowing water attempt to redirect river water to other locations nearby (Rivett et al., 2016; Karambelkar, 2017). Water-related rules in most nations now encompass not just prohibitions on the redirection of running water, but also several other rules, such as canal building, water consumption, hazardous material handling, water protection, and so on.

In some nations, for example, a landowner with a water brook/stream flowing through or beside his property is not permitted to freely use the water. Before constructing any diversion or other changes to the flowing water, he must first get permission. This holds for small towns, states, and the entire country.

The owner is also required by law to retain the water clean. This is true regardless of whether or not there is a hydroelectric plant on the water stream. If the stream has a hydroelectric plant, the landlord of the power plant is not permitted to minimize pollutants from in front of the dam wall and place it behind the wall.

3.4.2. Environmental Permits/Authorisations

Any owner of a hydroelectric facility must assess the project's environmental impact and adopt compensation measures to compensate for the power station's impact on the local ecosystem.

Legal issues have been a key stumbling block to the construction of hydroelectric capacity in various countries. Most countries, such as Germany and India, have taken the following legislative steps by enacting a "Federal Water Act" or "Hydro Policy" or equivalent legislation.

Throughout many countries, the law mandates a test technique to determine the environmental impact of development such as a hydroelectric power station. If the risk of environmental damage is too high, positive environmental actions must be implemented to compensate. The goal of such legal regulations is to provide consistency in the principles for measuring, analyzing, and describing the environmental impact of particular water-related initiatives. All governmental decisions touching the legitimacy must take into account the findings of the environmental impact assessment. On request, the relevant authority determines whether the study of environmental impact is required based on the proposed construction (Chandran, 2003; Mukherji, 2004). This evaluation will be made public.

A planner's documentation to the authorities must include, for example, a plan description, including the nature, location, and scope of the project, along with soil and ground needs. He must also explain strategies to reduce the negative environmental impact along with potential compensation actions. In addition, the planner must submit a summary of the environment and its components in the plan's affected region. He must also offer a summary of other very significant feasible solutions, including a comment about the critical grounds for choosing these ideas. Environmental consequences on neighboring countries must also be addressed during this process.

Stakeholder consultation is an important part of the planning and execution process of the project. Stakeholders including affected communities, regulators and other interested and affected parties are provided with pertinent information on the project such as the purpose of the proposed construction, potential negative and positive impacts as well as proposed management plans to adequately address environmental and social risks. The necessary documents are available for inspection. The authorities' decision is made public allowing stakeholders the opportunity to voice their opinions.

In India, similar to other countries, permits are required from the Ministry of Environment and Forests under the category "Irrigation, River Valley, Hydro-electric, and Multipurpose Projects',' prior to construction of a hydropower plant. . All options considered by the proponent of a project must be examined from an environmental perspective in the EIA (Environmental Impact Assessment). During the maintenance and process stages of a project, the plausible repercussions recognized for different aspects of the environment(aquatic, biological, hydrological, geological, socio-economical, terrestrial, fauna, and seismic) need to be briefly described for a various alternative. Thus, the ideas are provided for choosing the most reliable and best option based on environmental and social aspects and discarding additional options.

3.4.3. Actions of Inter-State Projects

The spatial orientation towards bodies of water is the specific criteria in water structure directives. Because a river may do distributed over two or more nations. On the other hand, the nation that is beneath the other, will not suffer any disadvantages. For instance, due to a territorially higher state, polluted river water is a problem. As a result, a unified regulation in neighboring states is obvious (Ibe & Onyema, 2013).

For instance, in Europe, the water framework directive, which has been in effect since 2000, is a European Union (EU) directive on water purification in member nations. The goal is to create a centralized water system that is sustainable environmentally and economically also for water usage. This could, for example, refer to the water quality and also the other issues that are addressed in the following sections. Thus, if a stream passes by several countries, approval is required before a hydropower plant can be operated.

3.4.4. Mutual Venture

Hydropower stations necessitate a significant amount of capital. Several advanced and emerging nations have state-run or private organizations that can finance power plants on their own. In some cases, mainly in developing nations, local governments do not even have adequate resources. So, in these circumstances, individual investors are allowed to promote partnerships with management to build plants. During these situations, the allies' interactions can be complex. An illustration from India that describes the aforementioned mechanism is provided below.

If it arrives at attracting more individual capital into India, strategy through domestic and foreign private companies and joint ventures between State Electricity Boards (SEB)/ the Public Sector Undertakings (PSU) is chosen. When a mutual venture company is formed, it becomes an autonomous body that is enrolled following the Companies Act and serves as a self-reliant builder. The joint venture contract among the two co-workers spells out each partner's level of involvement as well as the risks that will be shared during the project's construction and execution(Chandran, 2003; Mukherji, 2004).

Table 3.1. The Largest Three Hydroelectric Plants in India That Have Benefited From the CDM (as of May 2010)

UNFCCC Ref. No.	Title	State	MW capacity
1844	Budhil Hydro Electric Project, India (BHEP)	Himachal Pradesh	70
1326	Jorethang Loop Hydroelectric project	Sikkim	96
862	Allain Duhangan Hydroelectric project (ADHP)	Himachal Pradesh	192

3.4.5. Resettlement, Acquisition, and Rehabilitation of Land

The purchase of necessary government, private land, and forest necessitate lengthy plans, procedures and complex discussions by landowners who must sell their property. Contract issues arise as the consequence of, demands for jobs instead of land costs, land for land at the landowner's discretion, and so on for several projects. All of the designer's costs are regarded as project costs and they are permitted to be carried on through tariffs.

3.5. MECHANISM OF CLEAN DEVELOPMENT: INDIA AS AN EXAMPLE

According to the United Nations guidelines Core Convention on Climate Change (UNFCCC), both little and big hydropower stations are eligible to be recognized as clean development mechanisms and Mutual Implementation competitors. While following the Kyoto Protocol's compliance, to reduce emissions of greenhouse gases. If the size of the hydroelectric plant becomes lower than 15 MW, the project is deemed to be a small scale(Lewis, 2010). This is owing to variation in environmental impacts and it would also be bound by the methodology that has been approved AMS 1 D. If the capacity of a project is greater than 15 MW, it will be classified as a large scale project.

Table 3.2. Three of India's Smallest Hydroelectric Stations Have Benefited From the CDM

UNFCCC Ref. No.	Title	State MW capacity
1512	Deogad hydroelectric project in Maharashtra district Sindudurg, India by M/s Gadre Marine Export	Maharashtra 1.5
1566	Mini Hydel Scheme on Nagavali River, Andhra Pradesh	Andhra 1.7 Pradesh
662	Link Canal Mini Hydel project	Karnataka 1.5

Which will be directed to the ACM0002 established consolidated methodology. Under the scenario of storage hydropower station, the activity of a project is carried out in an actual reservoir to increase its capacity. If the activity of a project appears in the development of distinct reservoirs, the power station power output need to be higher than 4 W/m^2. So that, the project must follow the instructions of the World Commission on Dams (WCD) (Boyd et al., 2009).

REFERENCES

1. Ardizzon, G., Cavazzini, G., & Pavesi, G. (2014). A new generation of small hydro and pumped-hydro power plants: Advances and future challenges. *Renewable and Sustainable Energy Reviews, 31*, 746–761.
2. Armaroli, N., & Balzani, V. (2011). Towards an electricity-powered world. *Energy & Environmental Science, 4*(9), 3193–3222.
3. Aryandoust, A., & Lilliestam, J. (2017). The potential and usefulness of demand response to provide electricity system services. *Applied Energy, 204*, 749–766.
4. Bejarano, M. D., Sordo-Ward, A., Gabriel-Martin, I., & Garrote, L. (2019). Tradeoff between economic and environmental costs and benefits of hydropower production at run-of-river-diversion schemes under different environmental flows scenarios. *Journal of Hydrology, 572*, 790–804.
5. Boyd, E., Hultman, N., Roberts, J. T., Corbera, E., Cole, J., Bozmoski, A., ... & Liverman, D. M. (2009). Reforming the CDM for sustainable development: lessons learned and policy futures. *Environmental Science & Policy, 12*(7), 820–831.
6. Carrasco, J. M., Franquelo, L. G., Bialasiewicz, J. T., Galván, E., Portillo Guisado, R. C., Prats, M. M., ... & Moreno-Alfonso, N. (2006). Power-electronic systems for the grid integration of renewable energy sources: A survey. *IEEE Transactions on Industrial Electronics, 53(*4), 1002–1016.
7. Chandran, T. (2003). Investment in Power Sector: Past Trend and Short-Term Prospect. *Journal of Social and Economic Development, 5*(1), 40–68.
8. de Assis Espécie, M., de Carvalho, P. N., Pinheiro, M. F. B., Rosenthal, V. M., da Silva, L. A. F., de Carvalhaes Pinheiro, M. R., ... & dos Santos, F. N. G. A. (2019). Ecosystem services and renewable power generation: A preliminary literature review. *Renewable Energy, 140*, 39–51.
9. Fischer, D., Surmann, A., & Lindberg, K. B. (2020). Impact of emerging technologies on the electricity load profile of residential areas. *Energy and Buildings, 208*, 109614.
10. Georgiou, P., Tourkolias, C., & Diakoulaki, D. (2008). A roadmap for selecting host countries of wind energy projects in the framework of

the clean development mechanism. *Renewable and Sustainable Energy Reviews, 12*(3), 712–731.

11. Ghaneizadeh, A., Joodaki, M., Börcsök, J., Golmakani, A., & Mafinezhad, K. (2020). Analysis, design, and implementation of a new extremely ultrathin 2-D-isotropic flexible energy harvester using symmetric patch FSS. *IEEE Transactions on Microwave Theory and Techniques, 68*(6), 2108–2115.
12. Grandclément, C. (2019). Electricity engineers and the happening of behaviour: lessons from a real-scale experiment. *Distinktion: Journal of Social Theory, 20*(3), 246–263.
13. Guittet, M., Capezzali, M., Gaudard, L., Romerio, F., Vuille, F., & Avellan, F. (2016). Study of the drivers and asset management of pumped-storage power plants historical and geographical perspective. *Energy, 111*, 560–579.
14. Hanna, K., McGuigan, E., Noble, B., & Parkins, J. (2019). An analysis of the state of impact assessment research for low carbon power production: Building a better understanding of information and knowledge gaps. *Energy Research & Social Science, 50*, 116–128.
15. Hart, C., Watanabe, K., Song, K. J., & Li, X. (2007). East Asia Clean Development Mechanism: engaging East Asian countries in sustainable development and climate regulation through the CDM. *Geo. Int'l Envtl. L. Rev., 20*, 645.
16. Ibe, O. G., & Onyema, A. I. (2013). Concepts of reactive power control and voltage stability methods in power system network. *IOSR Journal of Computer Engineering, 11*(2), 15–25.
17. Jacobson, M. Z., & Delucchi, M. A. (2011). Providing all global energy with wind, water, and solar power, Part I: Technologies, energy resources, quantities and areas of infrastructure, and materials. *Energy Policy, 39*(3), 1154–1169.
18. Jiang, H., Lin, P., & Qiang, M. (2016). Public-opinion sentiment analysis for large hydro projects. *Journal of Construction Engineering and Management, 142*(2), 05015013.
19. Karakosta, C., Doukas, H., & Psarras, J. (2010). Technology transfer through climate change: setting a sustainable energy pattern. *Renewable and Sustainable Energy Reviews, 14*(6), 1546–1557.

20. Karambelkar, S. (2017). Hydropower Development in India: The Legal-Economic Design to Fuel Growth?. *Natural Resources Journal, 57*(2), 361–394.

21. Kuriqi, A., Pinheiro, A. N., Sordo-Ward, A., Bejarano, M. D., & Garrote, L. (2021). Ecological impacts of run-of-river hydropower plants—Current status and future prospects on the brink of energy transition. *Renewable and Sustainable Energy Reviews*, 110833.

22. Laghari, J. A., Mokhlis, H., Bakar, A. H. A., & Karimi, M. (2013). A new islanding detection technique for multiple mini hydro based on rate of change of reactive power and load connecting strategy. *Energy Conversion and Management, 76*, 215–224.

23. Lewis, J. I. (2010). The evolving role of carbon finance in promoting renewable energy development in China. *Energy Policy, 38*(6), 2875–2886.

24. Li, D. H., Yang, L., & Lam, J. C. (2012). Impact of climate change on energy use in the built environment in different climate zones–a review. *Energy, 42*(1), 103–112.

25. Manzano-Agugliaro, F., Taher, M., Zapata-Sierra, A., Juaidi, A., & Montoya, F. G. (2017). An overview of research and energy evolution for small hydropower in Europe. *Renewable and Sustainable Energy Reviews, 75*, 476–489.

26. Mukherji, R. (2004). Managing competition: Politics and the building of independent regulatory institutions. *India Review, 3*(4), 278–305.

27. Nyholm, E., Puranik, S., Mata, É., Odenberger, M., & Johnsson, F. (2016). Demand response potential of electrical space heating in Swedish single-family dwellings. *Building and Environment, 96*, 270–282.

28. Patzek, T. W., & Croft, G. D. (2010). A global coal production forecast with multi-Hubbert cycle analysis. *Energy, 35*(8), 3109–3122.

29. Reiche, D., & Bechberger, M. (2004). Policy differences in the promotion of renewable energies in the EU member states. *Energy Policy, 32*(7), 843–849.

30. Rivett, M. O., Cuthbert, M. O., Gamble, R., Connon, L. E., Pearson, A., Shepley, M. G., & Davis, J. (2016). Highway deicing salt dynamic runoff to surface water and subsequent infiltration to groundwater during severe UK winters. *Science of the Total Environment, 565*, 324–338.

31. Samanta, S., Rathore, A. K., & Thrimawithana, D. J. (2017). Analysis and design of current-fed half-bridge (C)(LC)–(LC) resonant topology for inductive wireless power transfer application. *IEEE Transactions on Industry Applications, 53(*4), 3917–3926.

32. Sharma, A. K., & Thakur, N. S. (2015). Resource potential and development of small hydro power projects in Jammu and Kashmir in the western Himalayan region: India. *Renewable and Sustainable Energy Reviews, 52*, 1354–1368.

33. Shen, J., Cheng, C., Cao, R., Shen, Q., Li, X., Wu, Y., & Zhou, B. (2019). Generation scheduling of a hydrodominated provincial system considering forecast errors of wind and solar power. *Journal of Water Resources Planning and Management, 145*(10), 04019043.

34. Shen, J., Cheng, C., Wang, S., Yuan, X., Sun, L., & Zhang, J. (2020). Multiobjective optimal operations for an interprovincial hydropower system considering peak-shaving demands. *Renewable and Sustainable Energy Reviews, 120*, 109617.

35. Singh, R. R., Chelliah, T. R., & Agarwal, P. (2014). Power electronics in hydro electric energy systems–A review. *Renewable and Sustainable Energy Reviews, 32*, 944–959.

36. Tang, N., Zhang, Y., Niu, Y., & Du, X. (2018). Solar energy curtailment in China: Status quo, reasons and solutions. *Renewable and Sustainable Energy Reviews, 97,* 509–528.

37. Truffer, B., Bratrich, C., Markard, J., Peter, A., Wüest, A., & Wehrli, B. (2003). Green Hydropower: The contribution of aquatic science research to the promotion of sustainable electricity. *Aquatic Sciences, 65*(2), 99–110.

Chapter 4

Physical and Technical Fundamentals of Hydropower Plants

CONTENTS

4.1. INTRODUCTION

Hydropower is an established technology having good reliability. The hydropower plant comprises constituents from mechanical, electrical, and civil engineering. Throughout operation and planning, normally it is vital to comprise information regarding hydraulics, hydrology, and environmental engineering with the knowledge of political and social problems, to discover an optimum solution (Sæþórsdóttir & Ólafsson, 2010; Gauché et al., 2017).

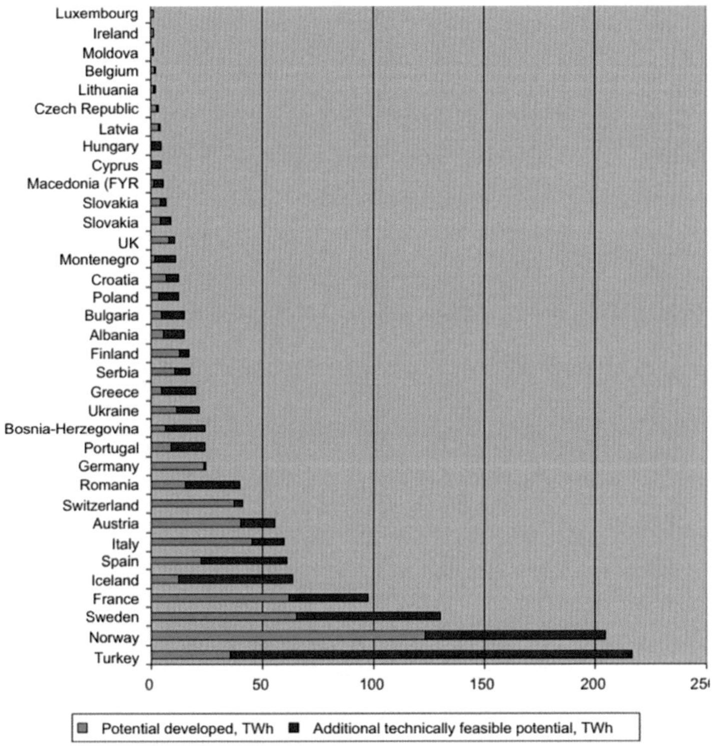

Figure 4.1. Established and leftover possible hydropower prospect in Europe.

[Source:https://www.yumpu.com/en/document/view/19557789/hydro-in-europe-powering-renewables-full-report-eurelectric.]

The hydropower plant usually comprises of the intake, a penstock, a waterway, a power station having mechanical and electrical equipments, and lastly the waterway to an outlet. It might not or might include the reservoir and a dam for the storage of water. The 3 key constituents of the equipment of Elmek are: (a) turbine, (b) generator, and (c) transformer. Additionally, there could be various other constituents like valves and gates, electronic equipments for governing the functionality of a station, switchyard, power cables, and the grid connections (Schägner et al., 2013; Ogino et al., 2019).

The plant of hydropower is almost tailored to use the accessible head and water, and several diverse kinds of turbines generally have been established: the utmost usual is the Francis and Pelton turbine for medium and high head circumstances and a Kaplan turbine used for the large flow and lower head systems.

The approaches of hydropower can approximately be categorized into 4 major kinds: RoR (run of river), storage (based on reservoir), PSH (pumped storage hydro), and the in-stream (hydro-kinetic) technologies. 6 kinds of hydropower approaches are given below.

4.1.1. Run of River (RoR)

These kinds of plants for hydropower primarily produce electric power from the accessible flow in the river. Small kind of storage might be comprised, permitting for alteration to the usage, however, the profile of generation will normally trail the profile of inflow.

4.1.2. Storage of Hydropower

Projects of hydropower projects having the reservoir can generally stock water for utilization later, usually by storing during the season of high flow (rainfall, spring) and discharging water stored during the season of low flow (winter). The reservoir provides higher adaptability and permits the plant to adjust to the profile of demand, seasonally and in a short term (days, hours).

4.1.3. Pumped Storage

A PSH scheme comprises the reversible plant of hydropower and 2 reservoirs, joined by the tunnel or a pipe. The prime objective is energy storage by pumping the water into an upper reservoir throughout periods of low requirement and produce power by discharging the water to the turbine throughout the periods of high requirement.

4.1.4. Hydro-Kinetic

This scheme isn't use and isn't develop much as compared to the other 3 schemes but provides the potential to prolong the variety of potential sites for the hydropower expansion to canals and rivers where it generally can utilize energy from the water flowing directly, instead of from the hydraulic heads made by the dams or some other governing structures.

4.1.5. Underground Hydropower Plants

Rock caverns and tunnels are vital elements of building in most of the large-scale hydro schemes, like tailrace and headrace tunnels, powerhouse, entrance tunnels, power cables shafts, and ventilation shafts or tunnels. In Norway, almost all of the large-scale hydro plants have usually been constructed underground ever since 1960 (Sæþórsdóttir & Ólafsson, 2010). If good quality rock is used, the majority of the penstocks, rock caverns, and tunnels can generally be utilized unlined, saving construction time and costly lining of concrete.

4.1.6. Small and Large Hydro Plants

It has normally become common to categorize the hydro plants either as "large" or "small," dependent on the capacity installed. There appears to be the certainty that small is attractive and that the small hydropower plants are eco-friendly. In few countries, small hydropower approaches are recognized for the expansion and get grants whereas larger schemes of hydropower aren't subsidized. If the effect per kW-h is observed, the representation becomes complex, and a single large plant of hydropower might simply have less effect as compared to several small plants having a similar capacity combined. This was discussed comprehensively in Gauché et al., (2017), in which the inference was that the utilization of cataloging with respect to their size must be circumvented, both since there isn't any clear association between impact and size and since the description "large" and "small" fluctuates broadly from one state to the other, as displayed in Table 15.4. Attempts have been made to categorize small plants as renewable and large plants as not renewable. This categorization is pure nonsense and must be circumvented. Large and small plants of hydropower are renewable equally, however, they might have dissimilar influences and be less or more ecological. It is thus good to talk over whether the scheme is justifiable and categorizes the project with respect to the index of sustainability (Scheffran & Battaglini, 2011; Oyewo et al., 2020).

4.2. LOCATING THE HYDROPOWER PLANT

Amongst the factors controlling the amount of production power and an overall performance of the hydro plant, location is very critical. There are various contemplations for classifying the most suitable location for small and large plant of hydropower.

4.2.1. Contemplations for Quantity of Water

When determining the location for a plant, usually it is essential to discover the average anticipated supply of available water for the production of power. As the number of rainfall changes on yearly basis and can't be anticipated very precisely, the judgment regarding the availability of water must be centered on the historical information of former eras. However still, there isn't any assurance for the yearly average volume of the water being accessible.

Because of changes in rain and temperature variations occasioning in changing amounts of the melting of snow, the release into the rivers changes on daily basis. Thus, the river plant of hydropower doesn't function constantly at the constant rate (Oyewo et al., 2020). The anticipated daily water flows during the year quite arbitrarily and are generally given in the shape of hydrographs for almost one complete year. To compute the required energy anticipated to be generated during a year, the mean everyday water flows must be arranged with respect to the water discharge into the river. This points to the curve of flow-duration, displaying the period of one year in which the particular flow is exceeded or reached as displayed in Figure 4.2.

Furthermore to the curve of flow-duration, several other curves must be well-thought-out, like the curve of mean-duration of the surface of water elevation in an upper reservoir dependent on output and input of water by the down flowing water.

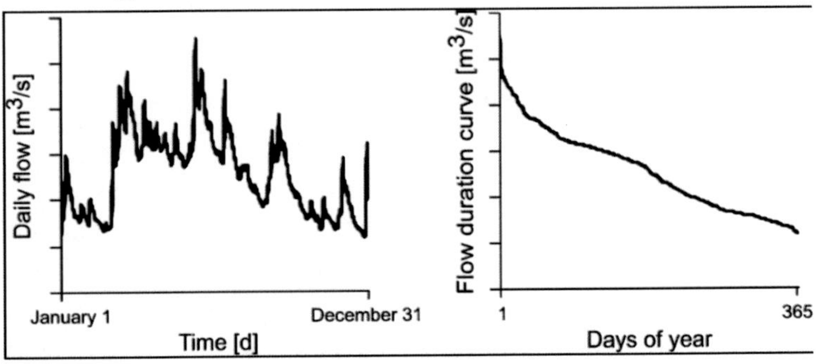

Figure 4.2. Hydrograph and the curve of flow-duration.

[Source: https://link.springer.com/chapter/10.1007/978–3-642–20709–9_3.]

When this data is considered the yearly output of energy can be computed by incorporating the equation of the power of water over an year. A better approximation of the anticipated mean water power production every year is possible only if the long-term records of streamflow are accessible, typically of twenty years. By utilizing stochastic hydrology with the methods of correlation, short accounts might be protracted to the longer appropriate period (Meyer, 2014; Kumar & Katoch, 2015).

When building the plant of hydroelectric power, it is essential to be aware of the anticipated average water accessibility. As the amount of rainfall isn't static and can't be anticipated, the empirical equation centered on the data of former decades is utilized. However, there isn't any assurance for the anticipated water availability. During the case of Rhine River, water quantity of the Rhine at the Rheinfelden in years 2003, 2001, and 1999, and the flow rates were recorded every day, as displayed in Figure 4.3. The River Rhine river has a yearly average water volume of nearly 1 thousand m^3s^{-1}. Whereas in the months of summer the average distribution amounts up to nearly 1,500 m^3s^{-1} and above, the months of winter deliver nearly 800 m^3s^{-1} of water. The period of 2003 was comparatively arid, and 2001 and 1999 were comparatively wet years. During the months of summer in the year 2003, when the climate is normally wet, the rate of flow was between 1,000–1,200 m^3s^{-1}. The months of October and January, conversely, exhibited even higher rates of flow as compared to the months of summer. In the year 1999, there prevailed 2 very wet phases, in February-March and the second one from May to July. Then, the flow rates accounted for more than four thousand m^3s^{-1}, which is normally a very high amount. The period of 2001 wasn't as wet compared to the year 1999 but yet displays high rates of flow from March to August.

Because of this instance, what logic does it contemplate to develop the plant centered on an average supply of water? If the plant isn't capable to generate electric power because of the water shortage, money is fruitless for offering redundant capacity.

Conversely, the accessibility of additional water is a chance missed to produce more energy that might bring income. The ultimate decision regarding capacity thus relies upon the water obtainability together with the interest rate, the market cost of electricity and consequential period of payback (Markoff & Cullen, 2008).

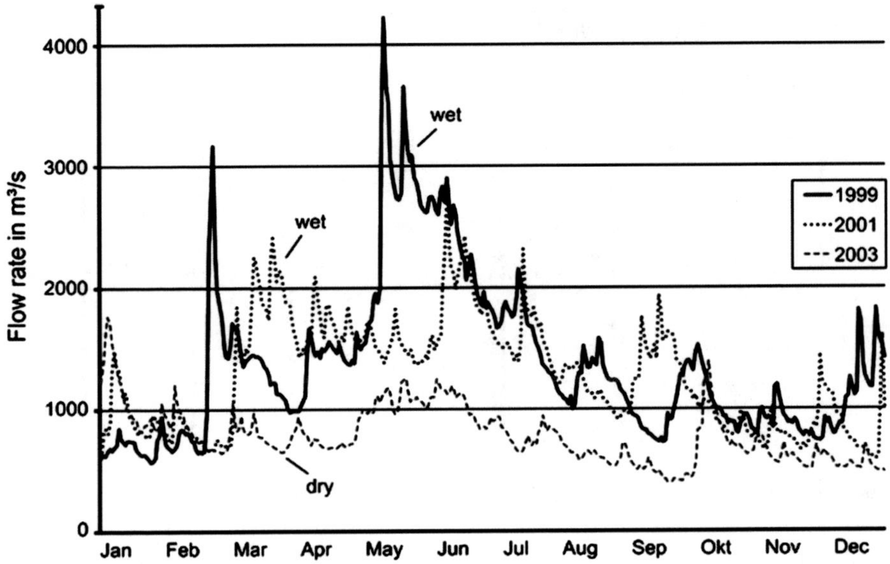

Figure 4.3. River Rhine's hydrograph in the state of Germany of two chosen rainy years (2001 and 1999) and one arid year (2003).

[Source: https://link.springer.com/book/10.1007/978–3-642–20709–9.]

One more specification that must be derived from water availability is secure electric power or firm power. This involves the approximation of the electricity amount that will be accessible for sure during the course of an entire year. For the approximation of secure electric power, the demanded amount of electric power must be accessible for a particular period over 1 year. Actually, though, the river flow rate might be lower or higher as compared to the turbine rate of flow, the flow needed by a turbine in order to generate maximum electric power according to its capability. When the rate of flow is quite lower, the turbine utilizes the water saved in a reservoir. When the rate of flow of a river is higher as compared to turbine's rate of flow, the reservoir is occupied with surplus water. Though, when a reservoir is occupied completely, and extra water is anticipated to reach, the surplus water is moved forward via the spillways of a dam (Hamlet & Lettenmaier, 1999).

4.2.2. Contemplations for Site of the Hydro Plant

Dependent on topographic circumstances at the location of a river where the plant is intended, there are numerous likelihoods to pinpoint the powerhouse.

One probable plan is to locate the powerhouse into the zero non-natural bay at the bank of a river.

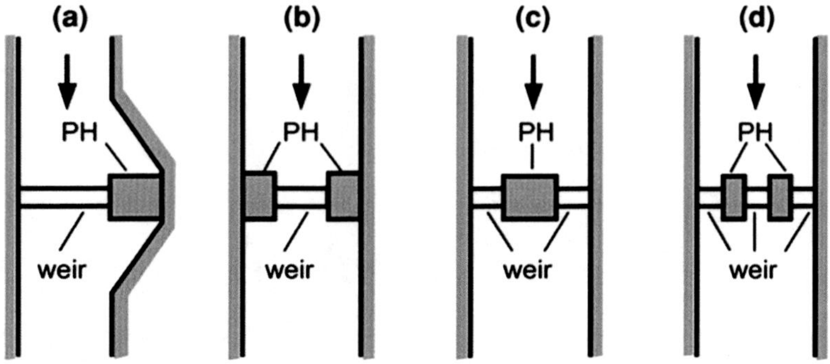

Figure 4.4. Design of weir and hydroplant in a bed of the river.

[ource: https://link.springer.com/chapter/10.1007/978–3-642–20709–9_3.]

Power house is the building of selection when the space comprising all supporting structures (sluice, weir) is quite large to settle amongst the banks of a river. These kinds of plants are known as "bay power plants." Several designs are displayed in Figure 4.4(a to d). The most significant benefit of arrangement (i) is the optimum route of overflows at a weir nearby to PH (powerhouse) (Solaun & Cerdá, 2019). If circumstances allow and the width of river is large then all of the structures comprising a weir, are positioned into the natural bed of river without disturbing the flood passage. In this situation, there are various opportunities to make the PH linking to a weir. Particularly when the entire plant is intended by two states as the mutual scheme and is positioned at the border of a river, alternative (ii) is appropriate. In this situation, every state can operate the power plant without snooping in function of the neighbor. Because of the difficult availability of PH, plan (iii) is suggested when foundation circumstances call for the position. A well-known organization is a possibility (iv) where each turbine together with a coupled generator is positioned into the isolated pier of a weir. Now in this situation, generally the units are reachable by the bridge passing the river.

Landscape and ecological characteristics occasionally lead to layouts contrary to the construction approaches mentioned above (Figure 4.4).

In the curves of a river, a power station is positioned at an outside bend due to the curved flow. In the rivers together with the transportation of sediment, the rotating current triggers corrosion of an outside bend and silting of inner parts of the bed of a river as displayed in Figure 4.5. Therefore, to shield the inner structure of PH from silting process, it must be positioned at an outer bank of the river (Bakker, 2013).

As a choice, in various cases, the diversion hydro plant might be well-thought-out which comprises of the canal before a weir takes water to the power plant, which don't disturb the key course of a river. The benefit of such kind is that the other events like transport by ships aren't affected by the existence of the power plant. Though, if the water quantity in a river isn't enormous, the key river might not have adequate water after the majority of water has usually been rerouted to a canal. The condition triggers issues for the events that before were reliant on the water of river. Though, after release from the plant, water has merged again into the key path, resulting in adequate flow of water again (Korten, 1987; Bakker, 2013).

Figure 4.5. Hydropower plant (100 year old) located in Black Forest Region of Germany, Right side shows the curved structure of a weir, Left side shows PH (powerhouse) with water channel.

[Source:https://www.researchgate.net/publication/298571181_Introduction_to_hydro_energy_systems_Basics_technology_and_operation.]

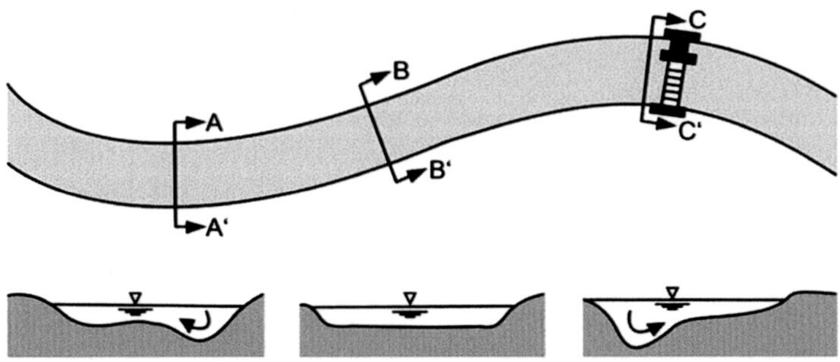

Figure 4.6. Development of silting and spiral flow in the bed of river.

[http://www.nzdl.org/cgi-bin/library?e=d-00000–00---off-0hdl--00–0----0–10–0---0---0direct-10---4-------0–1l--11-en-50---20-about---00–0-1–00–0--4----0–0-11–10–0utfZz-8–00&a=d&cl=CL1.14&d=HASH10e6c50b76a9b6493d7247.4.2.]

4.2.3. Manifold Reservoir System

In some of the situations, particularly in areas of mountains, rather than just 1 large area of catchment, there might be various smaller zones of catchment giving for normally more than 1 reservoir. The condition is displayed in Figure 4.7. The water from dispersed reservoirs can't gather to develop a large reservoir because of topographical restrictions, and the setting up of distinct plants for every trivial reservoir might not be feasible because of the absence of an appropriate location for setting up a turbine.

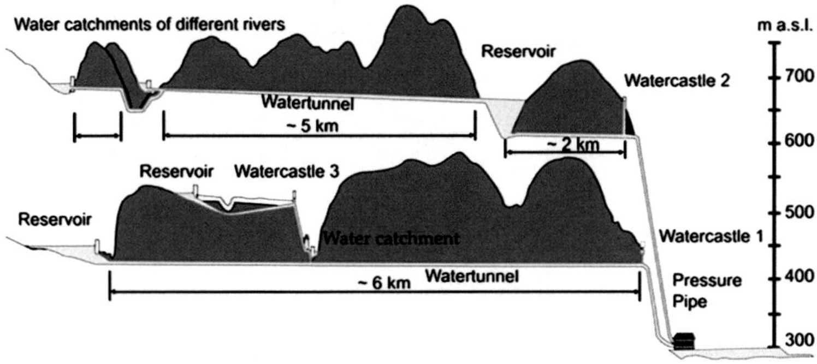

Figure 4.7. Manifold reservoir system of the storage hydropower plant located in Black Forest region of Germany.

[Source: https://link.springer.com/book/10.1007%2F978–3-642–20709–9.]

Figure 4.7 displays that the position on right hand side is an appropriate location for setting up a turbine, as it provides the biggest head for the production of power. In this situation, the water is transported to 1 mutual plant through the water tunnels moving on the ground and also through the trenches excavated via mountains. With the help of this planning, the reservoirs and turbine might be positioned several km apart and might not be detectable from one another (Depuru et al., 2011).

4.2.4. Cascaded Hydropower Plants

In particular cases, the large amount of water is accessible and the better water head is accessible over the large parallel distance. In a situation of the amalgamation, where the water quantity must be moved over extended distances, the common plant may not be an ideal option. Thus, to completely benefit from the quantity of water and fall height, a sequence of river hydro plants is made (Depuru et al., 2011). A representation of such kind of planning is displayed in Figure 4.8. It is essential to run the plants as the river plants and not as storing kind of plants, as water release from plant disturb the output of power from the consequent plants in series.

4.3. FUNDAMENTALS OF FLUID MECHANICS

4.3.1. Characteristics of Water

Water, a molecule comprising of two hydrogen atoms and one oxygen atom, possesses several extraordinary characteristics that are significant for the operation and construction of power plants.

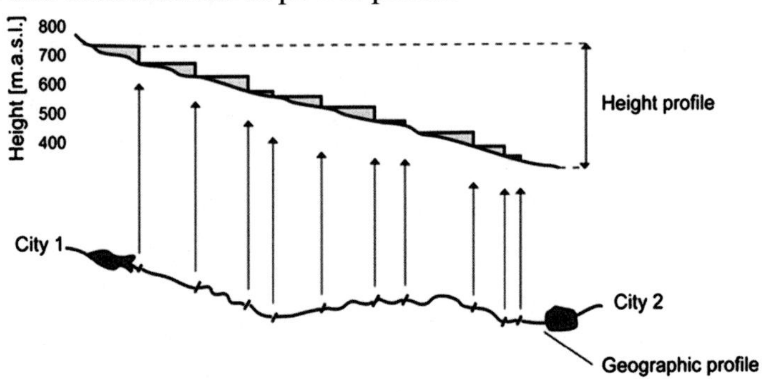

Figure 4.8. Representation of the cascaded hydropower plants on the River.

[Source: https://www.springer.com/gp/book/9783642207082.]

One feature of water is its density of 1 kg/L under usual circumstances, specifically, the temperature of twenty degrees C and the pressure of one bar. An additional feature is the water's incompressibility in the state of fluid. When the water pressure is upsurged, it passes on this pressure to the surrounds; water can't stock the impact of the upsurge in the pressure by itself getting compacted as for the gases state. For the power plants, this gives the meaning that the abrupt upsurge of pressure might affect the function and damage machine parts (Zinn et al., 2002; Nicot et al., 2009). To avert damages, several measures are taken to avoid abrupt pressure transmission from the parts of the machine of the hydropower plant.

Water displays an upsurge in volume if the temperature drops below 4°C. The maximum volume is reached at the temperature of –4°C. Because of such feature, water augments and might bring harm to the parts of machine. Thus, in the cold areas, freezing of the water should be avoided by appropriate systems of heating.

The temperature at which water evaporates, conversely, relies on pressure. The water evaporates at 100°C with the pressure of one bar, while the 20 millibar pressure makes the water to evaporate at 18°C.

In addition, water arrives into the solutions having salts, acid, and leaches. Because of this characteristic damage might be instigated.

4.3.2. Equation of Velocity

The flowing water, such as in the rivers, carries gravel and sand because of high energy water comprises. Sand begins travelling with water from the speed of around 0.30 ms^{-1}, smaller stones from around 1 ms^{-1}, and the bigger gravel from nearly 1.3 ms^{-1}. These elements travel with the flow of water till the water speed slows down at the curves of a river or after arriving at the reservoir. The impact of such moving is that gravel and sand gathered in piles ahead of the dam wall limit the function of the plant or trigger additional issues (Sedlmeier et al., 2011; Su et al., 2013).

Energy is kept in the water because of the fall height or elevation. When the waterfalls over the height, the potential energy is transformed into the corresponding kinetic energy as given in equation underneath:

$$mgh = \frac{1}{2} \cdot m \cdot v^2 = \frac{1}{2} \rho \cdot V \cdot v^2$$

where: m is the water falling mass; v is the velocity; g is the acceleration because of gravity; h is the height; q is the water density; and V is the volume.

and the equation can be amended in order to provide the velocity of water dropping over the height:

$$v = \sqrt{2 \cdot g \cdot h}$$

The equation recommends that this velocity upsurges with the square root of fall height. In order to give the imprint on the content of energy of the potential energy of water of 1 m^3 (cubic meter) of water at the level of one hundred meter comprises the energy of nearly 0.24 kWh.

4.3.3. Bernoulli's Equation

The running water in the river and stored water have two energy types: (i) kinetic energy because of the water flow and (ii) potential energy because of the height of water. The turbines of the hydroelectric plants are moved by the kinetic energy.

Similar to the further cases associated with energy, the flow of water also trails the energy conservation law. According to the theorem of Bernoulli, for the incompressible, non-sticky fluid in the stable flow, the totality of pressure, kinetic and potential energy over volume is continuous at some point. The equation of Bernoulli is the exceptional form of equation of Euler resultant along with the flow of fluid, and can generally be given, as displayed below:

$$\frac{v^2}{2} + \frac{p}{\rho} + gh = constant$$

where: v is the flow speed; p is the pressure; ρ is the water density; g is the acceleration due to gravity; and h is the fall height.

If the water is kept at the fall height in an open reservoir, then above Equation has just 1 constituent for the potential energy. Kinetic head and the pressure head are missing.

4.3.4. Equation of Power

Therefore, the entire power produced by water in the hydroelectric plant because of the height is generally expressed as (Lanzafame, 2015):

$$P = \eta \cdot \dot{m} \cdot g \cdot h = \eta \cdot \rho \cdot Q \cdot g \cdot h$$

where: P is the total power generated; $\dot{m}$ is the flow of mass of the falling water = Q. ρ total effectiveness of power stations; ρ is the water density; Q is the water rate of flow; g is the acceleration due to gravity; and h is the fall height.

The equation exhibits that the power output from the power plant is typically proportional to 2 parameters, explicitly height of fall and volume of flow. The succeeding vital parameter is total effectiveness, which can usually be enhanced through appropriate choice and function of the machinery. The total efficiency is a product of generator efficiency and turbine efficiency by ignoring the losses in the pipes. The efficiency of turbine normally ranges between 0.85 to 0.96, dependent on the design and type of turbine utilized, and considers losses of efficiency because of turbulence and friction amongst the entry of turbine and the exit of draft tube. The losses of friction inside a generator produce noise and heat in the powerhouse and the machinery and are incorporated in the efficiency of generator of nearly 98%. The inclusive efficiency of a station can be increased by upsurging the number of units fixed, particularly when the flows are varying (Dursun & Alboyaci, 2010; Rehman et al., 2015).

To acquire the high water head, the reservoir of water must be located very high, and the electric power production unit must be positioned very low. The height of the reservoir is concluded by topographical factors, like height of the river bed, quantity of the water, and some additional environmental factors. The position of the production unit of power can be accustomed as per amount of the power that is to be produced. Though, the position of power production unit is subjected to topographical limitations as normally the power production unit is set up at the lower levels as compared to local level of the ground to acquire the supreme water head.

The total rate of flow can generally be altered through the significant part of a hydropower station: a penstock. The dimensions of penstock must be adjusted for the assessed power of hydropower stations.

4.3.5. Continuity Equation

This equation recommends that because of the incompressible nature of water, the product of cross-section area of the flow and flow velocity remains persistent. This equation is expressed as follow:

$$v_1A_1 = v_2A_2 = v_3A_3 = \ldots = \text{constant}$$

where: v_3, v_2, v_1 are velocities at 3 different sections with areas A_3, A_2, A_1.

Table 4.1. Temperature at Which Water Evaporates at Dissimilar Pressures

Temperature (Celsius)	Pressure (bar)
24	$3.00 x 10^{-2}$
22	$2.64 x 10^{-2}$
20	$2.34 x 10^{-2}$
18	$2.06 x 10^{-2}$
16	$1.82 x 10^{-2}$
14	$1.60 x 10^{-2}$
12	$1.40 x 10^{-2}$
10	$1.23 x 10^{-2}$
8	$1.07 x 10^{-2}$
6	0.0093
4	0.0081
2	0.0071

The continuity equation exhibits that by the variation of flow cross-section, the variation in velocity of flow must happen to satisfy this equation. For the smaller cross-sections, the velocity of flow upsurges, and for the larger cross-section, it reduces (Gude, 2015).

4.3.6. Cavitation

Bearing in mind the effects of the equation of continuity together with the equation of Bernoulli recommends that with the reduction in the cross-section area, an upsurge in the velocity of flow occurs that upsurges the fluid's kinetic head. As the sum of all 3 terms in the equation of Bernoulli must remain persistent, this upsurge in the kinetic head outcomes in decrease of pressure if the water level doesn't change significantly.

Water possesses the feature of boiling at the low temperature with the pressure reduction, and if the static pressure reduces considerably, it causes evaporation. Due to evaporation, the water vapor bubbles formed are moved towards the zone with low pressure where these bubbles break down triggering the water rush in order to fill the vacant spaces of bubbles. If the event known as 'cavitation' occurs near the equipment surfaces like blades of turbine, is this much strong that it leads to everlasting damage of parts besides triggering the efficiency loss (Misyura, 2016).

Thus cavitation can generally be described as the creation of spaces inside the body of liquid flowing when the particles of moving liquid flop to obey the border of the passageway. Non-observance of liquid particles to obey the borders takes place when there is inadequate pressure in order to overwhelm the inertia of liquid particles and also to compel them to adopt the pathway with respect to the curve of flow.

Water, under usual circumstances, specifically, at the temperature of twenty degrees celsius and the pressure of 1 bar, is normally fluid. The evaporation of water with the similar pressure at hundred degrees celsius. With the reduction in pressure, evaporation of water occurs at lower temperature and this triggers problems with the power plants. With the pressure of nearly 20 millibars, evaporation of water occurs at 18 degrees celsisu as can be observed from Table 4.1.

Figure 4.9. Destruction of the blade of rotor as an outcome of the cavitation.

[Source:https://www.thermal-engineering.org/what-is-cavitation-damage-definition/.]

In the storage plants, for example, water should be lapped in from the lower levels in order to be forced up to an upper level of water. At the time, when the water is lapped in, the adjoining pressure is decreased. The outcome is that water evaporates at low temperature and the low pressure point. After being lapped in this water then reaches the zones at the pumping blades and pressure increases again in the pipes (Wei et al., 2011; Alkhalidi & Al-Jraba'ah, 2021). Because of higher pressure, the vapor of water precipitates and turns into fluid, in the shape of tiny drops. The tiny drops possess high velocity and they crash with the parts of machine, such as the turbine or the pump, with this velocity. Cavitation doesn't only look like when the water is lapped but also with the water in pipes.

4.4. ISSUES OF SUSTAINABILITY

Hydropower plant has numerous benefits than the other bases of electrical energy, particularly matched with the thermal power plant: hydropower is renewable, carbon-free, clean, and well-matched for incorporation into a grid, frequently backing the steadiness of the grid. But hydro plants can have negative influences, particularly on a native level, where weirs, dams, diversions, and variations in the flow of river might affect ecosystems and make problems (Al-Maamary et al., 2017; Hirsch et al., 2018). Weirs and dams create hurdles that might halt the migration of fish.

Table 4.2. Description of Small Hydro Plants by Fitted Capacity in Few Countries

Country	Small hydro plant described by fixed Capacity/MW
China	50.0
Brazil	30.0
India	25.0
Canada	<50.0
EU	20.0
Sweden	1.50
Norway	10.0
United States of America	5e100

Hydropower approaches often vary the regime of flow in the rivers and also the regime of levels of water in reservoirs and lakes and can thus have undesirable outcomes for biodiversity and ecosystems. Run of river projects normally don't vary the regime of flow, whereas storage hydro projects usually lead to variations in the yearly flow regime, for instance, by reducing flows during summer and upsurging winter flows. The transportation of sediments might be changed, creating issues in the intake pond and reservoir where the sediments are frequently accumulated and then downstream of a dam where decreased concentrations of sediment might lead to the upsurged riverbed and bank corrosion and reduced stability (Plate, 1993; Joskow, 2019).

The building of reservoirs and dams will cause loss of land, and occasionally, there is a necessity for the relocation of individuals who formerly lived in an area of the reservoir. All probable impacts must be examined during the phase of planning and contemplation in the assessment of benefit and cost, comprising not just financial values but the influence on society and the environment in a broader sense. But if influences can't be averted, it is normally likely to lessen few drawbacks if carried out during the phase of planning. Most of the undesirable impacts can usually be reduced and at times entirely evaded by cautious design, occasionally creating the win-win situation.

A decent summary of the environmental and social effects of hydro plant development is available in (Sæþórsdóttir & Ólafsson, 2010). The design of the hydro plant development in the majority of the countries has become time-consuming and complex, to comprise all dissimilar benefits. As an instance, displayed in Figure 4.10 is the procedure of licensing for the large hydro plants in the state of Norway.

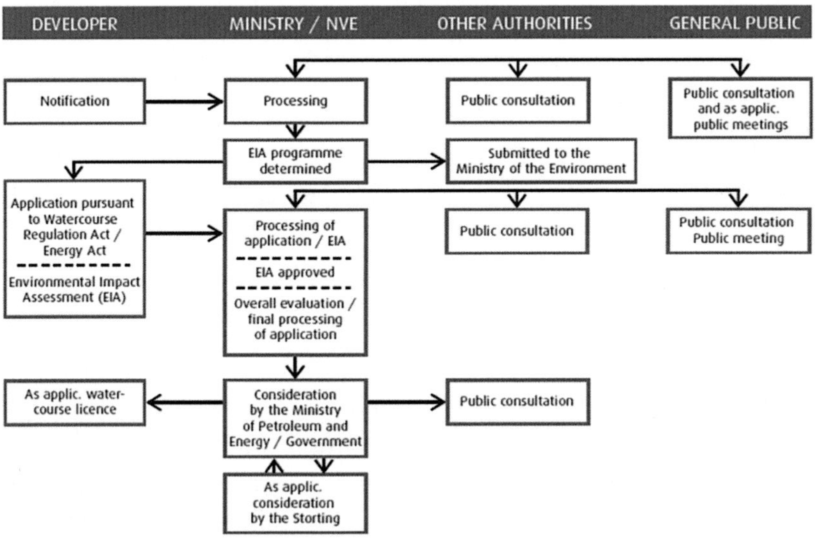

Figure 4.10. Licensing procedure for hydropower in Norway.

Source: Norwegian Ministry of Petroleum and Energy. (2008). The Legal Framework. (Internet).

[Source:https://www.regjeringen.no/globalassets/upload/oed/pdf_filer/fakta-heftet/evfakta08/evfacts08_kap04_eng.pdf.]

To evaluate the continuing feasibility of the plant, it is essential to comprehend all of the economic, environmental, and social concerns of the development. The problems of sustainability are becoming progressively significant for hydropower plant development, and IHA (International Hydropower Association) has established 3 tools of sustainability to control planning, execution, and the function of hydropower plant: IHA Sustainability Assessment Protocol, IHA Sustainability Guidelines, and Hydropower Sustainability Assessment Forum. The tools of sustainability can generally be acquired from the Web page of IHA (Amponsah et al., 2014). The IHA protocol comprises between 19–23 appropriate topics of sustainability, dependent on the phase of development of a plan. It is an outcome of the long process comprising several different shareholders: environmental and social NGOs, governments, commercial organizations, the hydropower sector, and the development banks, signified by IHA. Few topics comprised are biodiversity, infrastructure safety, indigenous people, resettlement, sedimentation and erosion, water quality, and downstream flow regimes. This protocol has generally been tested in almost 16 states.

4.4.1. Life-Cycle Assessment (LCA) of Hydropower Plants

The projects of hydropower possess a long life, almost 80 years (Soimakallio et al., 2011; Hertwich et al., 2015). To assess total benefits and costs, energy efficiency, environmental effects, and an impact on emissions of GHG (greenhouse gas), it is significant to contemplate the complete life-cycle for the project comprising examination, planning, building, function, maintenance, renovating, and ultimately retiring. The method of LCOE (Levelized Cost of Electricity) discoursed in the segment of cost is centered on the assessment of life-cycle of the economic accomplishment. Additionally, 2 other significant parameters of performance can also be stated: EPR (Energy Payback Ratio) and GHG (Greenhouse) emissions.

4.4.2. GHG Emissions

Most of the life cycle GHG emissions approximations for hydropower plant lies in the middle of about 4–14 g(CO_2eq) (kilowatt hour)^{1}, but under definite situations, there exists the threat of larger amounts of green house emissions (Hall et al., 2010). The usual range can generally be matched with 1 thousand g(CO_2eq) (kilowatt hour)^{1}for the plants of coal power, 500 for the plants of gas power, and 800 for the plants of oil power . Additional renewables also possess very low green house emissions, analogous or marginally higher as compared to hydropower (Figure 4.11).

4.4.3. Energy Payback Ratio (EPR)

EPR can be described as a ratio of the entire energy generated during the usual life span of a system divided by energy needed to construct, preserve, and fuel it. The higher ratio specifies better performance for envrioment. If this ratio of a system is near to unity, it ingests approximately the energy that is generated. Hydropower possesses the highest energy payback ratio of all electric power production technologies, having ratios varying from 171 to 268 for the run of river plants and 204 to 281 for the storage plants. It can generally be matched with usual ratios between 1.5 to 7.0 for fossil fuels, 19 to 33 for the wind turbines, and 14.1 to 16.2 for the plants of nuclear power (Figure 4.12).

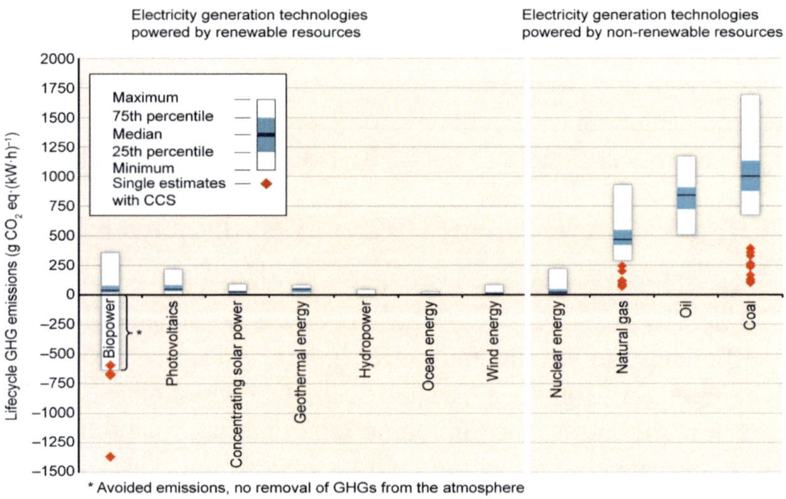

Figure 4.11. Approximations for the life-cycle global greenhouse gas (GHG) emissions in the generation of electricity.

[Sourcce: https://www.bafu.admin.ch/bafu/de/home.html.]

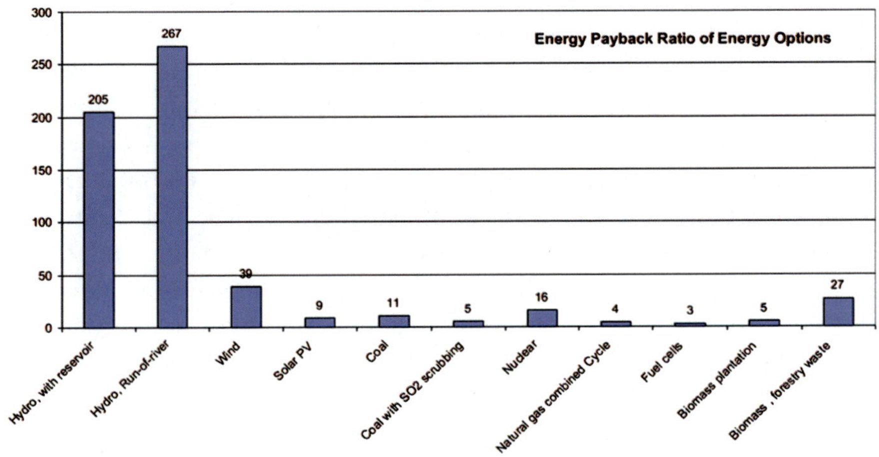

Figure 4.12. EPR (energy payback ratio) for some diverse electricity production technologies.

[Source: https://www.sciencedirect.com/science/article/abs/pii/S0301421508002401.]

4.4.4. Influence of Climatic Variations

The source prospects assessments given in Tables 4.1 to 4.3 have generally been centered on past hydrological circumstances and are thus effective just for the current climate. In the near future, with the varying climate, the prospects of hydropower could fluctuate because of the climatic variation effects on the sources of water that hydropower generation relies on:

- Variation in flow of river variability;
- Variations in volume of the flow of river;
- Variations in sediment loads that might affect the storage capacity of the reservoir;
- Variations in extreme events of flow (droughts, floods).

Numerous studies were appraised and briefed (Hall et al., 2010). The majority of studies have just concentrated on the variation effects in the volume of the flow of river, and some have comprehended and counted the impacts of the other 3 points.

In Kibert et al. (2000) the global and regional variations in the production of hydropower for the prevailing systems of hydropower were examined, centered on the worldwide evaluation of climatic variation and its impacts on the flow of the river by the year 2050 (Modahl et al., 2013). At higher latitude and in the majority of tropics, the models of climate typically envisage growing runoff and precipitation, whereas, in mid-latitudes, the estimate is that the rainfall will normally decrease. Areas with declining resources of water and reducing production of hydropower can be discovered near the Mediterranean, in Australia, in Southern Africa, and Western and the Central America. Areas having growing resources of water and growing production of hydropower can be discovered in Northern Europe, East Asia, East Africa, and Canada. The overall impact on hydropower and runoff was discovered to be trivial and possibly marginally positive (Soimakallio et al., 2011).

4.5. ISSUES OF COST

The price for the projects of hydropower is dependent on-site and can thus change considerably from project to project. The key constituents of cost are (a) up-front cost of investment, (b) cost of operation and maintenance, and (c) cost of decommissioning. LCOE comprises all these components of cost for the complete life of a project and is normally given in the units of US c (kilowatt hour)^{1}. The LCOE for hydro relies on these elements of cost and on (d) capability factor, (e) life of the project, and (f) capital expense.

Cost of capital comprises the price of civil constructions (tunnels, dams, PH), Elmek's equipment (gates, turbine, generators, transformer), powerlines, roads, planning and recompensing measures (resettlement, fish ladders). The cost of civil works is normally the largest portion for large developments, but also the smaller schemes, the price of Elmek can usually be larger (Amponsah et al., 2014). Usual expense of capital for the hydropower presently changes from <1000 US dollars (kilowatt)^{1} to 3000 US dollars (kilowatt)^{1} for the large hydropower and from 1550 to 6000 US dollars (kilowatt)^{1} or even higher for the smaller hydropower projects (Isaacman & Sneddon, 2000; Amponsah et al., 2014). There are several instances of projects having costs of 510 US dollars (kilowatt)^{1} for, particularly better sites. Usual life for hydro changes from 41 to 81 years (Hall et al., 2010).

Table 4.3 provides the calculated LCOE for the hydro for some usual amalgamations of the acute parameters: discount rate and cost of investment, for the life of 41 & 81 years. Operation and maintenance price was set to nearly 2.5% annually, and the factor of capacity was set at 45%. This provides LCOE in the range of 1.5 to 9.5 c (kWh) and most amalgamations group in the range 3.0 to 5.0 c (kilowatt hour) or 31 to 52 US dollar (Megawatt hour).

This generally can be matched with LCOE for some additional technologies, look at Table 4.4 and Figure 4.13. These are centered on statistics from numerous current projects, accumulated by the International Renewable Energy Agency (IRENA). Hydropower is an innovation with the lowermost average expense, trailed by on-shore wind, bioenergy, and geothermal. The closer observation into the IRENAs cost database discloses that small hydropower usually has a higher price as compared to larger one and also there are significant area dissimilarities. Areas like North America and Europe, where the larger portion of accessible resources have usually been established, exhibit the higher expense as compared to the regions with large unused potential, like South America and East Asia.

4.6. INCORPORATION INTO THE WIDER ENERGY SYSTEM

Moreover, to offering capacity and energy to fulfill electrical demands, hydropower plant has various features that allow it to offer additional services in order to make the operation of power system more trustworthy.

Table 4.3. LCOE Approximation for Parameters for short-term Projects of Hydropower from 2010 to 2020 (Sayboth, 2013)

Cost of investment (US dollars / kilowatt)	**Rate of Interest (%)**	**Operation & Maintenance cost (%/y)**	**Capacity factor (%)**	**Years of Lifetime**	**LCOE (US cents/ kilowatt hour)**	**Years of Lifetime**	**LCOE (US cents/ kilowatt hour)**
1,000	7.0	2.45	44	41	2.5	81	2.4
1,000	3.0	2.45	44	41	1.7	81	1.5
1,000	10.0	2.45	44	41	3.3	81	3.3
2,000	7.0	2.45	44	41	5.1	81	4.8
2,000	3.0	2.45	44	41	3.5	81	2.9
2,000	10.0	2.45	44	41	6.4	81	6.2
3,000	7.0	2.45	44	41	7.6	81	7.3
3,000	3.0	2.45	44	41	5.2	81	4.4
3,000	10.0	2.45	44	41	9.8	81	9.4

The plants of hydropower can begin producing electric power with very low startup costs and short notice; offer quick variations in production, and possesses high efficiency of part-load.

Table 4.4. Maximum, Average, and Minimum LCOE for Several Electricity Production Techniques in the Year 2016

	Bio-power	**Geo-thermal**	**Solar CSP**	**Solar PV**	**Solar PV**	**Fossil fuels**	**Hydro-power**	**Wind on-shore**	**Wind off-shore**
Maximum	17.0	11.3	31.2	27.9	27.9	14.0	24.6	14.1	20.8
Average	8.1	6.4	24.2	13.1	13.1	10.0	5.1	5.6	12.3
Minimum	6.1	4.3	18.2	5.3	5.3	4.5	1.8	2.4	9.6

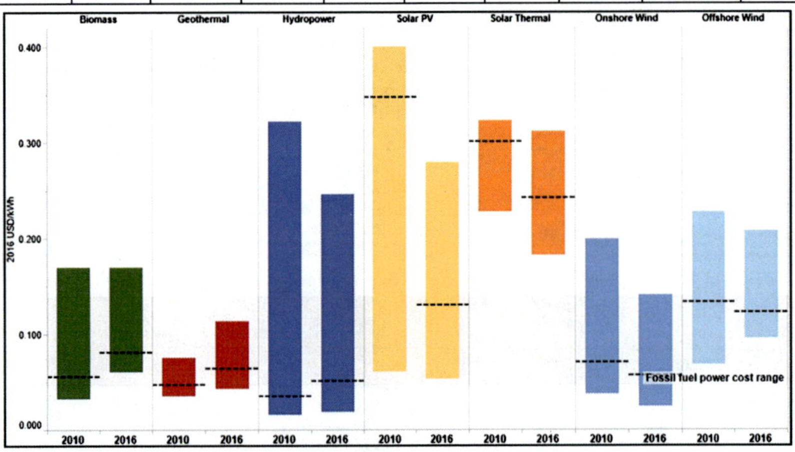

Figure 4.13. Trends in LCOE of electric power from several renewable technologies of utility scale. The dotted lines are overall average, bars specify a usual price range, and the gray area usual range of price for electric power from the fossil fuels.

[Sourcce: https://www.irena.org/statistics.]

The capability to quickly vary output in reaction to the requirements of system without experiencing large reductions in effectiveness makes the plants of hydropower appropriate to offering the harmonizing services known as regulation and the load following. The hydropower can offer the capacity to reinstate the power station to function without even depending on network electricity transmission network.

4.7. FUTURE DISTRIBUTION

Because just nearly 25% of the practical prospect and almost 50% of the economic prospect have been established, the potentials for additional developments are better. Numerous approximations of future distribution of hydropower have usually been made, nearly all envisage a significant increase worldwide.

The IPCC's AR4 supposed that hydropower might back to 17% of the overall electric power supply by the year 2030. Some of the hydropower industries envisage the even elevated market prospect of 8750 TWh a[1] to be touched by the year 2050 (Wei et al., 2011). The current investigation by IEA displays that hydro might double its involvement by the year 2050, touching 2 thousand GW of overall capability and around seven thousand TWh (Amponsah et al., 2014). The majority of this development would generally come from large hydro plants in evolving economies and emerging countries (Figure 4.14).

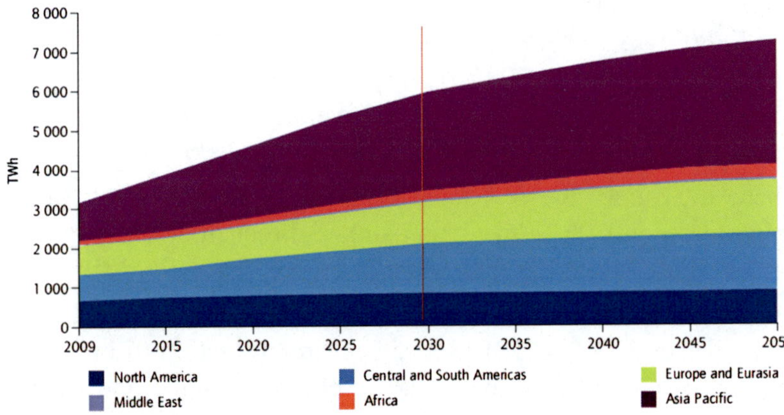

Figure 4.14. Hydropower Roadmap for the generation of hydroelectricity to the year 2050.

[Sourcce:https://cigreindia.org/CIGRE%20Lib/CIGRE-Paris-Session-2014-papers-Report/C1/C1_302_2014.pdf.]

REFERENCE

1. Alkhalidi, A., & Al-Jraba'ah, Y. K. (2021). Solar desalination tower, novel design, for power generation and water distillation using steam only as working fluid. *Desalination*, *500*, 114892.
2. Al-Maamary, H. M., Kazem, H. A., & Chaichan, M. T. (2017). The impact of oil price fluctuations on common renewable energies in GCC countries. *Renewable and Sustainable Energy Reviews*, *75*, 989–1007.
3. Amponsah, N. Y., Troldborg, M., Kington, B., Aalders, I., & Hough, R. L. (2014). Greenhouse gas emissions from renewable energy sources: A review of lifecycle considerations. *Renewable and Sustainable Energy Reviews*, *39*, 461–475.
4. Antikainen, J., Gebert, R., & Møller, U. (2011). Review of the greater Mekong sub-region regional power trade. *Sida Review 2011*, *9*.
5. Arango-Aramburo, S., Turner, S. W., Daenzer, K., Ríos-Ocampo, J. P., Hejazi, M. I., Kober, T., ... & van der Zwaan, B. (2019). Climate impacts on hydropower in Colombia: A multi-model assessment of power sector adaptation pathways. *Energy policy*, *128*, 179–188.
6. Bakker, K. (2013). Constructing 'public' water: The World Bank, urban water supply, and the biopolitics of development. *Environment and Planning D: Society and Space*, *31*(2), 280–300.
7. Beck, L., & Bernauer, T. (2011). How will combined changes in water demand and climate affect water availability in the Zambezi river basin?. *Global Environmental Change*, *21*(3), 1061–1072.
8. Chen, L., Stone, R., & Richardson, D. (2012). A study of mixture preparation and PM emissions using a direct injection engine fueled with stoichiometric gasoline/ethanol blends. *Fuel*, *96*, 120–130.
9. de Jong, P., Tanajura, C. A. S., Sánchez, A. S., Dargaville, R., Kiperstok, A., & Torres, E. A. (2018). Hydroelectric production from Brazil's São Francisco River could cease due to climate change and inter-annual variability. *Science of the Total Environment*, *634*, 1540–1553.
10. Depuru, S. S. S. R., Wang, L., & Devabhaktuni, V. (2011). Electricity theft: Overview, issues, prevention and a smart meter based approach to control theft. *Energy policy*, *39*(2), 1007–1015.
11. Devitt, P., & Hitchcock, R. K. (2010). Who drives resettlement? The case of Lesotho's Mohale Dam. *African Study Monographs*, *31*(2), 57–106.

12. Doso, O., & Gao, S. (2020). An overview of small hydro power development in India. *Europe, 251*, 643.
13. Dursun, B., & Alboyaci, B. (2010). The contribution of wind-hydro pumped storage systems in meeting Turkey's electric energy demand. *Renewable and Sustainable Energy Reviews, 14*(7), 1979–1988.
14. Ehrlich, P. R., & Holdren, J. P. (1971). Impact of population growth. *Science, 171*(3977), 1212–1217.
15. Emodi, N. V., Chaiechi, T., & Beg, A. R. A. (2019). The impact of climate variability and change on the energy system: A systematic scoping review. *Science of the Total Environment, 676*, 545–563.
16. Gauché, P., Rudman, J., Mabaso, M., Landman, W. A., von Backström, T. W., & Brent, A. C. (2017). System value and progress of CSP. *Solar Energy, 152*, 106–139.
17. Goldemberg, J. (2007). Energy choices toward a sustainable future. *Environment: Science and Policy for Sustainable Development, 49*(10), 7–17.
18. Gude, V. G. (2015). Energy storage for desalination processes powered by renewable energy and waste heat sources. *Applied Energy, 137*, 877–898.
19. Hadian, S., & Madani, K. (2015). A system of systems approach to energy sustainability assessment: Are all renewables really green?. *Ecological Indicators, 52*, 194–206.
20. Haimes, Y. Y., Matalas, N. C., Lambert, J. H., Jackson, B. A., & Fellows, J. F. (1998). Reducing vulnerability of water supply systems to attack. *Journal of Infrastructure Systems, 4*(4), 164–177.
21. Hall, P. J., Mirzaeian, M., Fletcher, S. I., Sillars, F. B., Rennie, A. J., Shitta-Bey, G. O., ... & Carter, R. (2010). Energy storage in electrochemical capacitors: designing functional materials to improve performance. *Energy & Environmental Science, 3*(9), 1238–1251.
22. Hamlet, A. F., & Lettenmaier, D. P. (1999). Effects of climate change on hydrology and water resources in the Columbia River Basin 1. *JAWRA Journal of the American Water Resources Association, 35*(6), 1597–1623.
23. Hamlet, A. F., Huppert, D., & Lettenmaier, D. P. (2002). Economic value of long-lead streamflow forecasts for Columbia River hydropower. *Journal of Water Resources Planning and Management, 128*(2), 91–101.

24. Hatch, C. E., Fisher, A. T., Ruehl, C. R., & Stemler, G. (2010). Spatial and temporal variations in streambed hydraulic conductivity quantified with time-series thermal methods. *Journal of Hydrology*, *389*(3–4), 276–288.
25. Hertwich, E. G., Gibon, T., Bouman, E. A., Arvesen, A., Suh, S., Heath, G. A., ... & Shi, L. (2015). Integrated life-cycle assessment of electricity-supply scenarios confirms global environmental benefit of low-carbon technologies. *Proceedings of the National Academy of Sciences*, *112*(20), 6277–6282.
26. Hirsch, A., Parag, Y., & Guerrero, J. (2018). Microgrids: A review of technologies, key drivers, and outstanding issues. *Renewable and sustainable Energy Reviews*, *90*, 402–411.
27. Hochstein, M. P., & Prebble, W. M. (2006). Major engineering constructions on top of a high-temperature geothermal system: problems encountered at Tokaanu, New Zealand. *Geothermics*, *35*(4), 428–447.
28. Isaacman, A., & Sneddon, C. (2000). Toward a social and environmental history of the building of Cahora Bassa Dam. *Journal of Southern African Studies*, *26*(4), 597–632.
29. Islam, M. A., Rahman, M. M., Heimann, K., Nabi, M. N., Ristovski, Z. D., Dowell, A., ... & Brown, R. J. (2015). Combustion analysis of microalgae methyl ester in a common rail direct injection diesel engine. *Fuel*, *143*, 351–360.
30. Joskow, P. L. (2019). Challenges for wholesale electricity markets with intermittent renewable generation at scale: the US experience. *Oxford Review of Economic Policy*, *35*(2), 291–331.
31. Juyal, N., Chamyal, L. S., Bhandari, S., Bhushan, R., & Singhvi, A. K. (2006). Continental record of the southwest monsoon during the last 130 ka: evidence from the southern margin of the Thar Desert, India. *Quaternary Science Reviews*, *25*(19–20), 2632–2650.
32. Kandlikar, S. G. (2001). A theoretical model to predict pool boiling CHF incorporating effects of contact angle and orientation. *J. Heat Transfer*, *123*(6), 1071–1079.
33. Khan, M. J., Bhuyan, G., Iqbal, M. T., & Quaicoe, J. E. (2009). Hydrokinetic energy conversion systems and assessment of horizontal and vertical axis turbines for river and tidal applications: A technology status review. *Applied Energy*, *86*(10), 1823–1835.

34. Kibert, C. J., Sendzimir, J., & Guy, B. (2000). Construction ecology and metabolism: natural system analogues for a sustainable built environment. *Construction Management & Economics*, *18*(8), 903–916.
35. Klemun, M. M., & Trancik, J. E. (2019). Timelines for mitigating the methane impacts of using natural gas for carbon dioxide abatement. *Environmental Research Letters*, *14*(12), 124069.
36. Korten, D. C. (1987). Third generation NGO strategies: A key to people-centered development. *World Development*, *15*, 145–159.
37. Kumar, D., & Katoch, S. S. (2015). Sustainability assessment and ranking of run of the river (RoR) hydropower projects using analytical hierarchy process (AHP): A study from Western Himalayan region of India. *Journal of Mountain Science*, *12*(5), 1315–1333.
38. Kuriqi, A., Pinheiro, A. N., Sordo-Ward, A., & Garrote, L. (2019). Flow regime aspects in determining environmental flows and maximising energy production at run-of-river hydropower plants. *Applied Energy*, *256*, 113980.
39. Lanzafame, G. (2015). Collective molecular dissipation on Navier–Stokes macroscopic scales: Accretion disc viscous modeling in SPH. *International Journal of Modern Physics D*, *24*(03), 1550028.
40. Léauthaud, C., Belaud, G., Duvail, S., Moussa, R., Grünberger, O., & Albergel, J. (2013). Characterizing floods in the poorly gauged wetlands of the Tana River Delta, Kenya, using a water balance model and satellite data. *Hydrology and Earth System Sciences*, *17*(8), 3059–3075.
41. Liang, J. Y., Zhang, J. L., Zhang, X., Yuan, S. F., & Yin, C. L. (2013). Energy management strategy for a parallel hybrid electric vehicle equipped with a battery/ultra-capacitor hybrid energy storage system. *Journal of Zhejiang University Science A*, *14*(8), 535–553.
42. Markoff, M. S., & Cullen, A. C. (2008). Impact of climate change on Pacific Northwest hydropower. *Climatic Change*, *87*(3), 451–469.
43. Meyer, A. H. (2014). Federal Regulatory Barriers to Grid-Deployed Energy Storage. *Colum. J. Envtl. L.*, *39*, 479.
44. Misyura, S. Y. (2016). Droplets boiling crisis of ethanol water solution on duralumin substrate with SiO2 nanoparticles coating. *Experimental Thermal and Fluid Science*, *75*, 43–53.

45. Modahl, I. S., Raadal, H. L., Gagnon, L., & Bakken, T. H. (2013). How methodological issues affect the energy indicator results for different electricity generation technologies. *Energy Policy*, *63*, 283–299.
46. Mote, P. W., Parson, E. A., Hamlet, A. F., Keeton, W. S., Lettenmaier, D., Mantua, N., ... & Snover, A. K. (2003). Preparing for climatic change: the water, salmon, and forests of the Pacific Northwest. *Climatic Change*, *61*(1), 45–88.
47. Nicot, J. P., Oldenburg, C. M., Bryant, S. L., & Hovorka, S. D. (2009). Pressure perturbations from geologic carbon sequestration: Area-of-review boundaries and borehole leakage driving forces. *Energy Procedia*, *1*(1), 47–54.
48. Ogino, K., Dash, S. K., & Nakayama, M. (2019). Change to hydropower development in Bhutan and Nepal. *Energy for Sustainable Development*, *50*, 1–17.
49. Oyewo, A. S., Aghahosseini, A., Ram, M., & Breyer, C. (2020). Transition towards decarbonized power systems and its socio-economic impacts in West Africa. *Renewable Energy*, *154*, 1092–1112.
50. Phan, H. T., Caney, N., Marty, P., Colasson, S., & Gavillet, J. (2009). Surface wettability control by nanocoating: the effects on pool boiling heat transfer and nucleation mechanism. *International Journal of Heat and Mass Transfer*, *52*(23–24), 5459–5471.
51. Plate, E. J. (1993). Sustainable development of water resources: a challenge to science and engineering. *Water International*, *18*(2), 84–94.
52. Pledger, A. G., Rice, S. P., & Millett, J. (2014). Reduced bed material stability and increased bedload transport caused by foraging fish: a flume study with juvenile Barbel (Barbus barbus). *Earth Surface Processes and Landforms*, *39*(11), 1500–1513.
53. Rathi, R., Prakash, C., Singh, S., Krolczyk, G., & Pruncu, C. I. (2020). Measurement and analysis of wind energy potential using fuzzy based hybrid MADM approach. *Energy Reports*, *6*, 228–237.
54. Rehman, S., Al-Hadhrami, L. M., & Alam, M. M. (2015). Pumped hydro energy storage system: A technological review. *Renewable and Sustainable Energy Reviews*, *44*, 586–598.
55. Rosas, M. A., Vanacker, V., Viveen, W., Gutierrez, R. R., & Huggel, C. (2020). The potential impact of climate variability on siltation of Andean reservoirs. *Journal of Hydrology*, *581*, 124396.

56. Sæþórsdóttir, A. D., & Ólafsson, R. (2010). Nature tourism assessment in the Icelandic Master Plan for geothermal and hydropower development. Part I: rapid evaluation of nature tourism resources. *Journal of Heritage Tourism*, *5*(4), 311–331.
57. Schägner, J. P., Brander, L., Maes, J., & Hartje, V. (2013). Mapping ecosystem services' values: Current practice and future prospects. *Ecosystem Services*, *4*, 33–46.
58. Scheffran, J., & Battaglini, A. (2011). Climate and conflicts: the security risks of global warming. *Regional Environmental Change*, *11*(1), 27–39.
59. Sedlmeier, F., Horinek, D., & Netz, R. R. (2011). Spatial correlations of density and structural fluctuations in liquid water: A comparative simulation study. *Journal of the American Chemical Society*, *133*(5), 1391–1398.
60. Seyboth, K. (2013). Intergovernmental panel on climate change (IPCC). *Encyclopedia of Energy, Natural Resource, and Environmental Economics, 1, 1-24.*
61. Shrestha, S., Khatiwada, M., Babel, M. S., & Parajuli, K. (2014). Impact of climate change on river flow and hydropower production in Kulekhani hydropower project of Nepal. *Environmental Processes*, *1*(3), 231–250.
62. Soimakallio, S., Kiviluoma, J., & Saikku, L. (2011). The complexity and challenges of determining GHG (greenhouse gas) emissions from grid electricity consumption and conservation in LCA (life cycle assessment)–A methodological review. *Energy*, *36*(12), 6705–6713.
63. Solaun, K., & Cerdá, E. (2019). Climate change impacts on renewable energy generation. A review of quantitative projections. *Renewable and Sustainable Energy Reviews*, *116*, 109415.
64. Spalding-Fecher, R., Joyce, B., & Winkler, H. (2017). Climate change and hydropower in the Southern African Power Pool and Zambezi River Basin: System-wide impacts and policy implications. *Energy Policy*, *103*, 84–97.
65. Su, J., Ong, R. C., Wang, P., Chung, T. S., Helmer, B. J., & De Wit, J. S. (2013). Advanced FO membranes from newly synthesized CAP polymer for wastewater reclamation through an integrated FO-MD hybrid system. *AIChE Journal*, *59*(4), 1245–1254.

66. Tena, A., Batalla, R. J., Vericat, D., & López-Tarazón, J. A. (2011). Suspended sediment dynamics in a large regulated river over a 10-year period (the lower Ebro, NE Iberian Peninsula). *Geomorphology*, *125*(1), 73–84.
67. Wei, J. H., Pan, L. M., Chen, D. Q., Zhang, H., Xu, J. J., & Huang, Y. P. (2011). Numerical simulation of bubble behaviors in subcooled flow boiling under swing motion. *Nuclear Engineering and Design*, *241*(8), 2898–2908.
68. Weisser, D. (2007). A guide to life-cycle greenhouse gas (GHG) emissions from electric supply technologies. *Energy*, *32*(9), 1543–1559.
69. Wilhite, D. A., & Glantz, M. H. (1985). Understanding: the drought phenomenon: the role of definitions. *Water International*, *10*(3), 111–120.
70. Wu, F. F., & Varaiya, P. (1999). Coordinated multilateral trades for electric power networks: theory and implementation. *International Journal of Electrical Power & Energy Systems*, *21*(2), 75–102.
71. Zinn, R. A., Owens, F. N., & Ware, R. A. (2002). Flaking corn: processing mechanics, quality standards, and impacts on energy availability and performance of feedlot cattle. *Journal of animal Science*, *80*(5), 1145–1156.

Chapter 5

Economics of Hydropower

CONTENTS

5.1. INTRODUCTION

The major production of renewable energy is the current generated by hydropower. Hydropower produces almost 20% of the world's electricity. Almost 50% of electricity production of 1/3rd of every country has dependency on hydropower (in 2001) (Harrison et al., 2003; Førsund, 2005). Productivity, volume and share of electricity of top 10 greatest generators of hydropower (in 2009) is shown in the table below. China has only largest main production plant as it is the main producer of electricity produced by hydropower; In 2008, construction of the Three Gorge Dam was completed. China's 22% of current productivity is signified by water. Northway is the 6th main creator of power produced by water and 98% of its electricity need, is fulfill by hydropower whilst USA has only 6% sharing in the production of hydro powered electricity which is the least input in its production. Different use of hydropower is indicated by the different ratios of energy production and capacity between countries as peak load and base load capacity, and dissimilarities in the comparative accessibility of influxes (Hirsch et al., 2014; Xue & Sun, 2018).

Table 5.1. Highest Hydroelectric Generators in 2009

Country	**Annual generation of hydroelectric (TWh)**	**Capacity (GWh)**	**Installed % capacity of total capacity**
Republic of China	652	197	22
Canada	370	89	61
Brazil	364	69	86
United States (USA)	251	80	6
Russia	167	45	18
Norway	141	28	98
India	116	34	16
Venezuela	86	15	69
Japan	69	27	7
Sweden	66	16	44

[Source: https://en.wikipedia.org/wiki/List_of_largest_hydroelectric_power_stations.]

5.1.1. Environmental Problems

Hydropower is frequently called green energy as it does not produce injurious discharges. While the major problem to environment is the misuse of hydropower plant. Natural atmospheres or populated zones flood by artificially manufactured lakes called reservoirs. This is in contrast to, for example, the distant highlands of Norway, where reservoirs are created by natural lakes. Moreover, water extracted from watercourses are transferred over lengthy distances by tunnels. Lastly, pipelines may extend within mountains in tunnels clear have a direct path to the turbine from reservoir. Therefore, the natural environment is used for the hydropower system, limiting or negating the use of recreational natural facilities such as lakes, waterfalls and rivers (Abbasi, 2000).

Currently variations in the reservoir level and the quantity of water downstream of the facility has resulted in negative impacts on aquatic ecosystems,. The valuable freshwater fish like trout and salmon in the season of reproduction may suffer. In the regions of the former natural streams, lakes and rivers the variation in microclimate creates problem for agriculture along with the issues for aquatic life by altering the level of reservoirs. There are records of instances where the construction of dams results in the formation of methane gases when large underwater plant- or tree-covers are either released in processing of water in turbines or through direct releases (McCartney et al., 2000). When the dam is filled again, the rotting process continues as the dam banks can re-produce plants in dry seasons.

5.1.2. Fundamental Physics of Hydropower

By induction, electricity production by turbine driven by water is the basis of hydropower. The falling of water on the turbine from altitude and the initial energy is provided by gravity. Hydropower can be built on water taken from reservoirs which can include up to various years' value of influx, on free river streams, and dams with inadequate storing volume higher than the normal flow. When there is a seasonal influxes to fill the reservoir up and difference between the patterns of usage of current, the monetary use of hydropower is reservoir (Mosonyi, 1988; Schaefli et al, 2007). The concerning production and inflows in Norway in a normal year is shown in Figure 5.1.

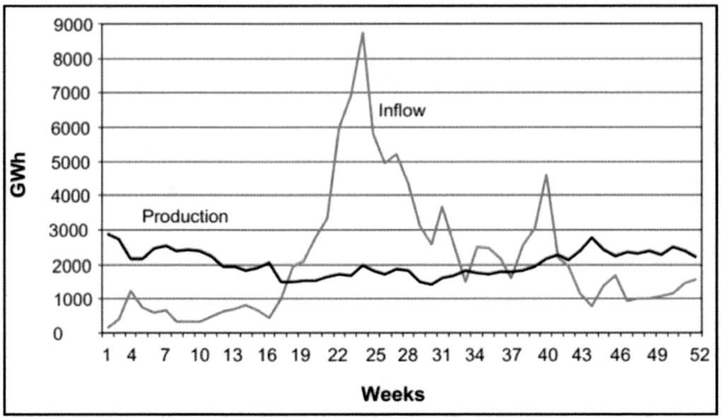

Figure 5.1. Weekly inflow and production of hydropower in Norway (2003).

[Source: https://www.uio.no/studier/emner/sv/oekonomi/ECON4930/v11/undervisningsmateriale/Hydropower%20economics4.pdf.]

The water from the initial autumn days to the late spring, summer, late autumn, and winter days is transferred through dams . In Norway when there is melting of snow in summer and spring as in other countries the reservoirs are filled up.

The generation function of a hydro plant represents the conversion of water for a plant into electricity with a reservoir

$$e_t^H = \frac{1}{a} r_t, \qquad (1)$$

In a period, r_t, e_t^H is the energy produced, in which the water is discharegd in this periods on the turbine, and a is the unit constraint or the input coefficient of water; illustration on number of units of water are mandatory to generate 1 kWh of power. In the power plant, the flow which is founded on river movement and there is no reservoir then it is replaced for the release of water. There will be impartiality in the relation of generation above for an effectively run process.

The altitude difference to the turbine level from the dam level the basis of current potential of per unit water. The current production changes when water releases and there will be a changed in reservoir level. The way of transportation of water away from turbine and permits new water to come in effects the current production. For an ideal flow of water, turbine is built. Power output per unit of water may be decreased by rather higher or lower inflow of water. According to use of a turbine there is the efficiency of atleast

80% and at most 95–96% of difference of energy conversion, and if more water is release to the turbine, then a reduction occurs again (Benitez et al., 2008; Mulat & Moges, 2014).

5.1.3. Engineering Cost Estimation

There is neither the entrance of real capital nor other existing contributions like material and labor. Although the capital can be repressed in an examination of handling the given capacity but it should have to deliver current generation ability. The capital structure represents the technology. The engineering strategy of optimum water release on to the turbines and the existed technology of feeding tunnels will be reflected by fabrication coefficient. Thorough engineering information about energy efficiency will be ignored by us.

For optimum management of power operations, the costs nature is important. Current operations should be affected only by variable costs, although capacities are fixed and present. Though, the water is the only input in our description and it does not buy from a market. The old flexible costs, i.e., the prices which changes with the level of productivity may be ignored as irrelevant and it is identified by empirical information. There is the same staff to look after the processes although there is a change in output. Current output level is not the function of maintenance as it is the function of size of capital structure. (However, the number of start-ups has dependance for wear and tear of turbines.) There will be zero current costs in our supposition. For hydropower this is a very accurate supposition. An opportunity cost is the form of variable cost which embodies water as stated above, i.e., using water in future is the profit of present cost. The decreased conversion of electricity is disregarded in the generation function, because of the abridged altitude the water falls. For the Norwegian system, this is a satisfactory simplification at our level of combination with huge changes in altitude in turbine plants and dams (typical height is higher than 200 meters) and with relatively few river stations. The use of the reservoir influences the coefficient and it is stated in more technically-oriented analyses. There is tremendous simplification in production function. The calculation of the input coefficient contains topology and the built wall because it gives height (Bekchanov et al., 2015).

Hydropower related two diverse decision problems are followed as the decision related to the running the operation of existing capacities and the investment choice in novel capacities. The chapter will only be focused on former problem. Advanced bathtub diagrams explain the main conclusions. Journal papers contain a lot amount of engineering literature, but there is a little literature on economics basis (Boasson et al., 2009; Ming et al., 2014).

5.2. THE BASIC MODEL FOR MANAGEMENT OF HYDROPOWER SYSTEMS

5.2.1. Hydropower with a Reservoir

There will be a statement of the simplest model that operates hydropower system. Distinct time will be used. This is the case for all applied use of the kind of model being examined. Flow and stock will be the types of variables. The date of the stock variables should be noted, for example, at the start or end of a period (Kodikara et al., 2010; Niu et al., 2018). The interpretation of movement variables would be presented as scales linked to understanding in a period.

The emptying and filling of the reservoir are the basis of water management. The water accumulation equation represents these activities

$$R_t \leq R_{t-1} + w_t - r_t,\ t = 1,..,T \qquad (2')$$

Quantity of water present in the reservoir at the completion of period t, is R_t. the influx towards the reservoir T *is* w_t that is the number of periods we are considering, and the discharge of water towards the turbines is r_t as explained in linking with the generation function. The quantity of water in the reservoir at the completion of period t is lower or equivalent to the quantity of water expected from the earlier period $t - 1$, along with the influx in the period and deducted the usage of water for generation of in the period.

There is no clearly description for evaporation from the reservoir (Zhou et al., 2020). For warmer countries this is define in the definition of inflow but it is fairly rational for a northern country like Norway. There is size limitation for reservoir

$$R_t \leq \overline{R} \qquad (3)$$

where $\overline{R}$ represents highest quantity the reservoir may contain, so overspill may occur. In Equation (2') there is severe dissimilarity in that case and $R_t = \overline{R}$. There is zero lower limit for ease, so $\overline{R}$ is maximal amount of water that can be used.

We can substitute the generation function (1) into the accumulation of water Equation (2') giving

$$R_t \leq R_{t-1} + w_t - r_t = R_{t-1} + w_t - ae_t^H \Rightarrow$$

$$\frac{R_t}{a} \leq \frac{R_{t-1}}{a} + \frac{w_t}{a} - e_t^H \qquad (4)$$

All the variables will be calculated in the unit of energy kWh to simplify the representation and rewrite the water buildup as

$$R_t \leq R_{t-1} + w_t - e_t^H,\ t = 1,..,T \qquad (2)$$

kWh is the measuring unit of all variables, but still there will be the discussion about utilization of water, etc., meanwhile this has a more instinctive demand.

5.2.2. The Social Planning Problem

In order to manage the capacities of the system over time, we have to frame an objective function for the problem. As reservoirs exist so our problem is dynamic so we need to choose the duration of water usage. For each period's requirement, the minimization of costs is the key engineering approach for meeting the demands (Morillo et al., 2020). Though, hydropower with zero fixed costs, and flexible costs is ideal to utilize the current hydropower system and it is not included in management problem. This will be the case in our assumption. Expanding the advantage of customers and generator from generating current is the simplest economic objective function. The objective function may be represented as

$$\sum_{t=1}^{T} \int_{z=0}^{e_t^H} p_t(z)\,dz \qquad (5)$$

The area under the demand functions $p_{t\,=}\,p_t\,(e_t^H)$ for every period is used to calculate advantage so termed customers and generator surplus as there is no variable costs (Attia et al., 2014).

The price idea in the study is in the form of communally best prices. Either these prices match with values that user truly pay or market prices, is separate debate. The objective function (5) has measuring unit of money, expresses a kind of communal welfare function irrespective of all distributional problems because a value finds the users' peripheral readiness to pay (Deb, 1999). The problem of the optimum administration of hydro capacities can be viewed in segregation, discounting any relations with the independent economy. A partial problem that is an approximation to a more general analysis is the real issue to deal with. Using the objective function, totality of producer and consumer surplus considering the limited reservoir and the water accumulation function, the social problem is optimized.

$$\max \sum_{t=1}^{T} \int_{z=0}^{e_t^H} p_t(z)\,dz \qquad (6)$$

subject to

$$R_t \leq R_{t-1} + w_t - e_t^H$$

$$R_t \leq \bar{R}$$

$$R_t, e_t^H \geq 0 \ , \ t = 1,..,T$$

$$T, w_t, R_o, \bar{R} \text{ given, } R_T \text{ free}$$

Some explanations are to be considered. Initially, no omitting is representing the neutral function of Equation (6). From 1 to 5 years, for a hydro management issue, the planning horizon T is in use. There will be an insignificant effect by any not counting a real interest rate of size 4–6 %, on the results (Attia et al., 2014). Additionally, at the completion of the planning horizon, the quantity of water is free which is left in the reservoir, usually indicating that nothing will be left. In the model formulation, these characteristics are subtle to vary.

The issue of ptimization is a distinct time dynamic programming issue, and for this type of problem special solution procedures have been established. However, there is no need to go into technical details, but it is necessary to understand the basics of qualitative outcomes by direct perceptive initially and then familiarize graphical demonstrations.

There are few rational economic suppositions. A supposition of a firmly optimistic quantity of electricity generated in every period appears apparent because concern is about a system-wide model. Moreover, it is rational to consider the social values which will be firmly optimistic for every period (Morillo et al., 2020). The social solution may not be better because there is no appearance of any time preferences and electricity is a perfect homogeneous good, but it can be made better if it is probable to monitor flawless arbitrage and maintain the equal price in each period. Consider a situation in which the worth for a period is greater than for all other times having a usual price, ever greater price than any of the other times. This means that by generating more electricity in this period the value of the function may be enhanced at the expense of the other periods. If all prices are equal, the value of the objective function may not be greater than that value. The feasibility of such a solution may be prevented by the restricted in the optimization problem (6).

For the issue of dynamic programming, Bellman's back induction is the general principle for solution. Therefore, the solution can be derived for the latter period, T. It tracks openly that the quantity of water accessible in this period is $R_{T\text{-}1} + w_T$. Then, we have that all the accessible water will be generated due to our supposition of no limit on the quantity of water to vacate at the completion of the planning period, yielding

$$e_T^H = R_{T-1} + w_T \quad (7)$$

The social price becomes

$$p_T = p_T(e_T^H) = p_T(R_{T-1} + w_T) \quad (8)$$

We see that the price and optimal amount of electricity are dependent upon the quantity of water moved from period T-1 to period T. The limit of this movement is

$$R_{T-1} \in [0, \bar{R}] \quad (9)$$

Hence, there are two corner solutions, when the largest quantity R is left to the following period and when the least quantity of zero is moved i.e., wholw accessible water is utilized in period T–1 keeping the reservoir unfilled. The quantity is in between these limits for an interior solution to the next period. Following basic possibilities are followed for a important perception of optimum pricing: assume one of the limitations on the reservoir is obligatory, an optimal price will change. Norwegian electricity regulator already stated this in 1968. The thinking makes it evident that the values maintains to the same level as far as an optimal solution is inact. The optimum quantity to leave to period T is $R_{T?1}$ and the quantity of water expected from the earlier period T–2 is $R_{T\text{-}2}$ going back in time to T-1 leaving the optimal quantity of electricity to be

$$e_{T-1}^H = R_{T-2} - R_{T-1} + w_{T-1} \quad (10)$$

It is known that in period T-1 the price would be the similar of period supposing the interior solution of the quantity $R_{T\text{-}1}$ to leave for period T, implying

$$p_{T-1}(e_{T-1}^{H}) = p_T(R_{T-1} + w_T) \Rightarrow e_{T-1}^{H} = p_{T-1}^{-1}(p_T(R_{T-1} + w_T)) \qquad (11)$$

Using (10) yields

$$R_{T-2} = R_{T-1} - w_{T-1} + e_{T-1}^{H} = R_{T-1} - w_{T-1} + p_{T-1}^{-1}(p_T(R_{T-1} + w_T)), \qquad (12)$$

where obtain the last expression we have substituted from Equation (11).

Going on to period T-2 we get

$$R_{T-3} = R_{T-2} - w_{T-2} + e_{T-2}^{H} = R_{T-2} - w_{T-2} + p_{T-2}^{-1}(p_T(R_{T-1} + w_T)) \qquad (13)$$

Hence maintaining the supposition that every optimum transmissions of water from a certain period to another are interior solutions, restating this rational right to period 1 regressive in time achiveved by consecutively introducing the solution for the level fo reservoir opening with $R_{T\text{-}2}$ suggesting that all the prices are the same in Equation (12), we obtain at last

$$R_0 = R_{T-1} - \sum_{i=1}^{T-1} w_i + \sum_{i=1}^{T-1} p_i^{-1}(p_T(R_{T-1} + w_T)) \qquad (14)$$

The final quantity of water $R_{T?1}$ which is left at the terminal period T is the only unknown in this equation. Using Equation (14), once solution is provided for this quantity all the other solutions for the quantities of electricity for every period and common price for every period track from Equations (7), (8), (12), etc.

Yet the possibilities of a corner solution will be considered. Assume that the quantity of water gone to the following period is t+1 going backward on the axis of timett he initial period we meet a corner solution where it can be optimum to vacate the reservoir, i.e., $R_{t+1} = 0$. By

Equation (14) we get

$$R_{t+1} = R_{T-1} - \sum_{i=t+2}^{T-1} w_i + \sum_{i=t+2}^{T-1} p_i^{-1}(p_T(R_{T-1} + w_T)) \Rightarrow R_{T-1} = \sum_{i=t+2}^{T-1} w_i - \sum_{i=t+2}^{T-1} p_i^{-1}(p_T(R_{T-1} + w_T)) \qquad (15)$$

The value was same from T to t+2; $p_{t+1} \geq p_{T,}$ must be lesser than the optimal value in period t+1 because in another case it will not be optimum to vacate the reservoir in the period t+1.

Assume that for parting a filled reservoir R in period s to period s+1 having

$s+1 < t$, we have the other corner solution, but for periods back from t to $s+1$ we have interior solutions for periods, period $t+1$ can be used in the part of the terminal period suggesting the value is same on given time interval and proceed as above to obtain

$$R_s = R_t - \sum_{i=s+1}^{t} w_i + \sum_{i=s+1}^{t} p_i^{-1}(p_{t+1}(R_t + w_{t+1})) \Rightarrow R_t = \bar{R} + \sum_{i=s+1}^{t} w_i - \sum_{i=s+1}^{t} p_i^{-1}(p_{t+1}(R_t + w_{t+1})) \quad (16)$$

Supposing that for all periods there are interior solutions from period s-1 back to the initial period there is a new value system with usually a lesser value than attained for periods from t+1, $p_u \leq +1 = p_u +1 = p_t +1$ (u=1,..,s), as unless the value in the coming period is greater exiting a filled reservoir demonstrating a danger of overspill will not be optimum. In the same way as above the level of the final (going backward value system) is established by solving for $R_{s\text{-}1}$ from

$$R_0 = R_{s-1} - \sum_{i=1}^{s-1} w_i + \sum_{i=1}^{s-1} p_i^{-1}(p_s(R_{s-1} + w_s)) \quad (17)$$

It was established that the value changes are visible because of the basic limitations of reservoir becoming active. During a yearly cycle, there may be numerous price changes depending on inflow conditions and demand. But it is rather evident that the changes we see must have additional explanatory factors in hydro-dominated courtiers like Norway.

5.2.3. The Bathtub Diagram for Two Periods

Variables from two successive periods are observed by noticing the optimization issue structure in the equation (6) producing dynamics, water buildup Equation (2). This means that the chief characteristics of the general solution may be captured by the sequence of two-period diagrams. There can be three circumstances for value-setting rules as we have explored above, the reservoir running empty, the reservoir continuing inside its bounds of vacant and the reservoir streaming full (Castelletti et al., 2010; Attia et al., 2014). To make this practical, focus on just two periods is enough. In a bathtub diagram, the solution for two periods can be demonstrated. Figure 5.2, displaying the total accessible water for the two periods as demand curves attached on every wall and the floor of the bathtub. There is the introduction of maximum storage. Influx in period 2 is *CD* and inflow and the primary water R_o in period 1 is *AC*. *BC* is the maximal storage. For period 1 the storing is calculated from *C* near the axis due to the choice of amount water to move to period 2 is done in period 1.X

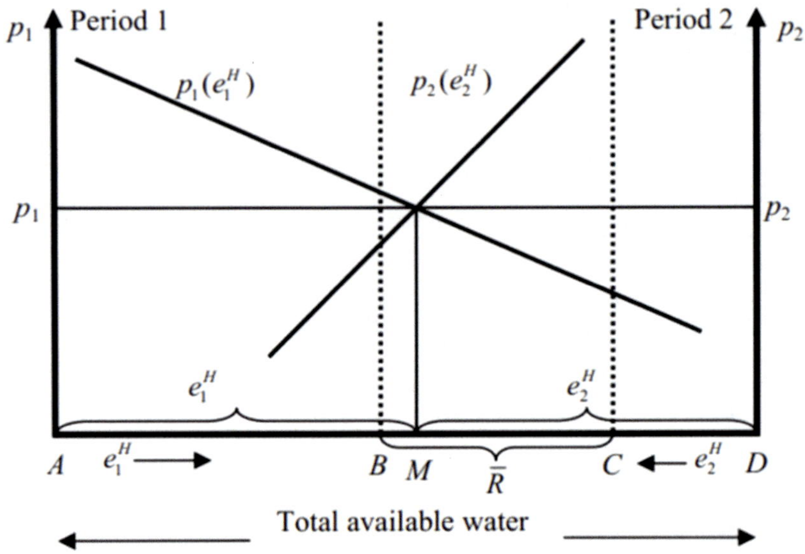

Figure 5.2. Illustration of the two-period bathtub with non-binding reservoir limitations.

[Source: https://www.uio.no/studier/emner/sv/oekonomi/ECON4930/v11/undervisningsmateriale/Hydropower%20economics4.]

The shared value for the two consecutive periods can be determined by the crossing of the demand arcs. The dispersal of electricity generation in two periods is shown on the floor of bathtub by point *M*. The ideal transmission shows the condition where the limit of reservoir is not touched, however, there is a shortage in the period 2 because all accessible water, *MC* + *CD*, in this period is consumed. Hence, *MC* is moved and saved to period 2 and the amount *AM* is used up in period 1. A positive value for both the periods is indicated by the total amount available consumed in both periods. Inside the perpendicular lines from *B* and *C*, the crossing of the demand arcs occurs signifying the maximal storable amount. The water price become same as the value for both periods, as water utilized in the period 1 is at the outflow of possible use in this period. From *B*, the demand arcs can cross to the left of the fragmented perpendicular reservoir capacity line as demonstrated in Figure 5.3.

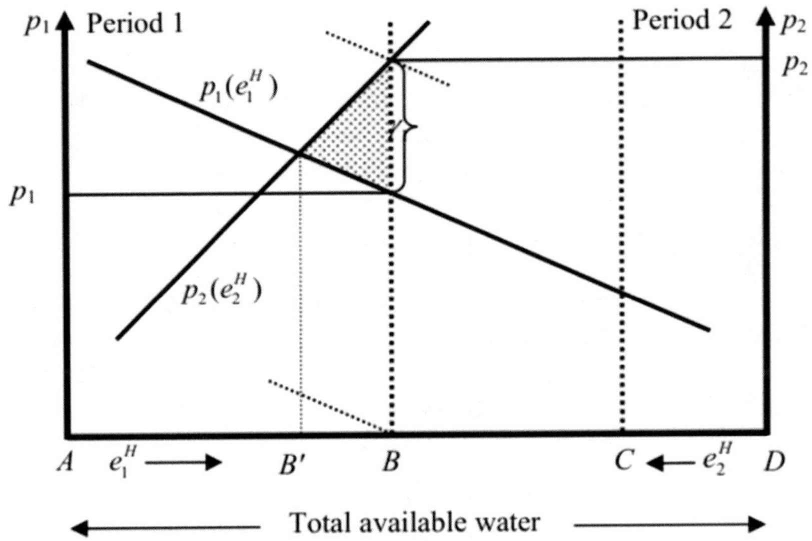

Figure 5.3. Social optimal with higher reservoir limitation binding in period 1.

[Source: https://www.uio.no/studier/emner/sv/oekonomi/ECON4930/v11/undervisningsmateriale/Hydropower%20economics4.]

As the value of water is higher in the subsequent period, the optimal distribution is now to collect the most quantity *BC* in period 1, and water that cannot be collected should be consumed, *AB*, in period 1. It may not be optimum with any leak in period 1, because of the supposition of non-satiation demand. As anticipated from the discussion below, the price is now greater in the following period (16). In this period the reservoir, comprising an influx of *CD* falling in the period and *BC* from the first period, is emptied. We go to the periods with scarcity from the period of the danger of overspill. Notice that for an extensive variety of period 1 demand arcs maintaining the similar period 2-arc, or vice versa, water allocation will be the same (Castelletti et al., 2010; Attia et al., 2014). The period 1- curve can be moved up, going from the level aimed at the period 2 water price and down to passing through *B* as designated by the spotted lines as substitute demand arcs. The value change in the periods can congruently vary significantly. The price of the objective function turn into lesser as implied by a binding reservoir constraint. Consuming the unrestricted solution as a standard, the noticeable triangle is decrease in entire producer and user additional because of the restricted magnitude of the reservoir, designated by the perpendicular spotted mark from B' to the crossing of demand arcs. For just two periods, the bathtub diagram may be used as in Figure 5.2 and 3, but for two consecutive periods, it may also be used within a multiperiod analysis. An arrangement of optimum solutions consuming two-period bathtub diagrams is illustrated by the two-period feature of the dynamics of the method.

5.2.4. Additional Constraints

There is a need to introduce an upper limit on the electricity generation capacity during a period, to enhance the realism of our model. A hydropower plant that has a mounted power capacity provides a limitation on the potential quantity of energy in a period when used maximally:

$$e_t^H \leq \bar{e}^H , \ t = 1,..,T \qquad (18)$$

The quantity of electricity that the structure can endure is also limited by the transfer system connecting consumers and producers. The limiting factor is often the transmission system. Since we are modeling at a system level, Equation (18) can also be used to signify this event. The same strategy of tackling the reservoir limitation in the earlier section can be used to take care of this constraint (Kelman et al., 1998; Madani & Mariño, 2009). A binding restraint on generation changes the prices upwards is a rather straightforward statement.

Sudden variations in the utilization of water may alter other usage of rivers or dams or may have negative environmental effects. Therefore, constraints are typically imposed on ramping rates.

5.2.5. Multiple Producers

A country can have multiple hydropower plants. In Norway, for example, there are 700 hydropower stations, and reservoirs form part of a majority of these plants. Therefore, it is interesting to determine under what conditions working with a combined scheme can be of significance. The same suppositions in Section 2 are considered here and the limitation on the reservoirs will be discussed in present section. One reservoir is allotted to each plant. The plants function autonomously, and a transmission system is not detailed, such as there may not be "hydraulic couplings" as they would be in plants with the unchanged river scheme. A significant outcome of discounting power, transmission, or production restrictions for each plant is in a way that during a single period, a plant can vacant its reservoir. It may be termed as *the perfect maneuverability* (Madani & Mariño, 2009; Ringkjøb et al,, 2018). But there is no such assumption that influxes may be directed to some reservoir. The influxes are plant or reservoir-definite. In the production functions, the plants have in overall varying construction coefficients, and the plant-specific fabrication coefficient deflates the water-buildup equation of the kind (2) for every plant, considering no wastage of water in generation. The total amount used up is a aggregating production from every plant, but the development issue remain the same as in Equation (6).

For the *aggregate* generation, distinctive solutions have been attained in every period, however, not essentially solutions on individual plants for the distribution of this generation. Assuming that every reservoir is vacated in the terminal period *T* and that demand is not satiate utilizing the backw induction standard, the total production of electricity is same as the total consumption due to the free terminal condition, and the value mutual to every unit is measured in the similar manner as clarified for the combined system considered above. In absence of any plant emptying its reservoir the value should be common to all plants and same as for period *T.* Same output can be achieved when going backward to period 1. However, there is a need to check the possibilities in presence of a corner solutions for individual plants and reservoirs and how the system value can vary (Laughton et al., 2000). The relationship between total supply and demand can change the system price, and there is no value for the prices of the individual plant. However, from the aggregate comportment of the system, the consumption trend for single plant does not essentially follow distinctively. The equilibrium quantity and piece can be determined by the interplay of whole supply and demand in a period *t*. However, for the optimal solutions, it may not matter in general which plants add to whole supply and what amount, as far as, when the system price changes, the plants have the optimum quantities in the reservoirs.

It can be concluded that the administration of single plant is not wholy measured in periods when the social value keep consistent. Contingent upon the magnitude of the reservoir comparative to the quantity of influx, plants differs from the total quantity of periods it allows to feed the reservoir. During a constant price regime, a plant having lesser storing capacity can vacate the reservoir numerous times covering various periods. An significant message to notice is that every plant should have a filled reservoir to transmit to the period with a value increasing uphill, where there is a system value variation, and every plant should have vacant reservoirs before the period if the price goes down for the solution to be optimal. Back in 1968, the director of the Norwegian electricity regulator expressed that 'not a single reservoir is empty before all are empty,' and no single reservoir is overflowing before all reservoirs are filled up (Laughton et al., 2000; Ringkjøb et al., 2018).

Though, if the presence of plants is considered with huge reservoirs, for instance, multi-year reservoirs, and for each plant, extra constraints are presented independently, then this combination estimation will not hold precisely. The estimation is only effective as an estimate will be the result of decreasing maneuverability of the reservoirs to various degrees, which will be more or less crude.

5.3. COSTS AND BENEFITS

A hydropower plant has rather diverse economics compared with any other power plant because numerous concerns including river navigation, irrigation, and water supply are involved along with consistent economic characteristics of cost of produced power (Birol et al., 2006; Singh & Nachtnebel, 2016). In actuality, some of these aspects, such as the impact of recreation facilities or irrigation are hard to quantify. Therefore, true economic study of a hydropower plant is a mix of qualitative and quantitative methods, especially for a large hydropower plant.

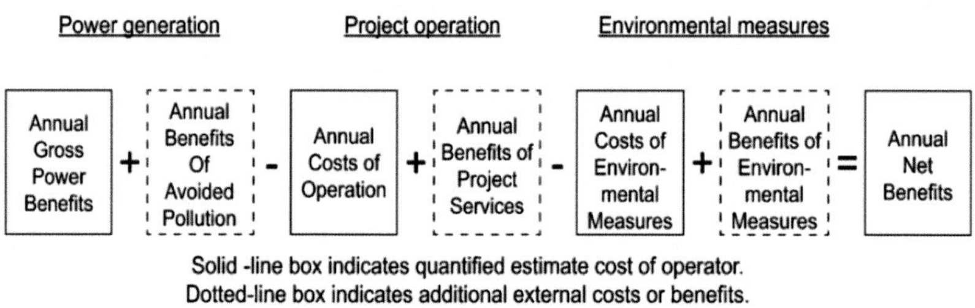

Figure 5.4. Overview of economic analysis.

[Source: https://link.springer.com/chapter/10.1007/98–3-642–20709–9_7.]

The main cost and profit components for approximating the net annual profits from a hydropower are presented in Figure 5.4 and explained in this section (Brown et al., 2018):

- Costs of operation: These kinds of costs show estimated future reinvestment costs, investment costs for the project, and current maintenance and operation costs.

- Costs of environmental measures: Various licensing choices present operating circumstances intended to mitigate damages, protect and improve environmental quality. From the perspective of the owner of the hydropower station, these changes may result in reduced power values and/or direct costs. For example, some direct costs linked with the building of fish passage services. In the same way, the flow of water is sometimes restricted because of the environmental actions to shelter flora and fauna that may reduce power production either because they change power production from times in which energy values are high to times in which energy values are low or they produce direct losses in accessible flow.
- Gross power benefits: these profits include the revenue from the trade of power or saving from the price of power if power has been occupied from a pricier source when the hydropower plant did not exist.
- Profits of environmental actions: Change in least flow requirements or fish screens, can improve recreational places, make wildlife and fish resources better and improve other features of environmental value. They are normally denoted as "non-power" benefits because these profits are dissimilar from the direct income from the trade of power.
- Benefits of avoided pollution: Comparative to substitute forms of power production, i.e., a coal-fired plant, hydropower generation creates fewer greenhouse gases or air pollution. The avoided pollution is reflected as an advantage of hydropower plants.
- Benefits of project services: hydroelectric projects can provide other benefits beyond power generation, i.e., water supply, flood control, river navigability, and economical prosperity of the region, and improvement of infrastructure

Details regarding the sale of power and the cost structure are given in the following sections.

5.3.1. Initial Costs

Since each location may present a distinctive set of problems, the budget of hydropower stations can barely be indiscriminate, such as essential investments in infrastructure, effort in moving equipment due to poor road infrastructure, the tunnels and the lengths of pipes, dissimilar geology, that are shown by the rate of the plant and thereafter on the generated power cost. Depending on the size of location and country, the primary expenses of hydropower plants are typically differ between 1500–6,000 Euro per kW. The capital cost was found to be about 1,300 Euro per kW, 30 years ago, in the study of the Itaipu hydropower plant, while according to press releases, in the Three Gorges Dam plant it was 1,000 Euro per kW. The capital budget per kW is greater than these values when comparatively smaller capacities are planned (Pilavachi et al., 2009).

In the case of greenfield plans, civil engineering works usually take 65–75% of the total primary cost, where no dam has occurred before the hydropower scheme, and meeting legal and environmental standards needs 15–20%.

The price of the project equipment i.e., generators, turbines, and the control systems an take about 10% of the whole cost. During operation, there are no fuel costs linked with hydropower plants; hence maintenance and operation costs are rather low in contrast to a fossil power plants.

In cases where dams already exist, i.e., in the case of nongreen field projects for some other purposes such as flood control or recreation; the locations, at comparatively lower costs, can be upgraded to have hydropower plants. In the same way, as equated with reservoir-type power stations, run-of-river might also be least expensive, as fewer civil workings are considered.

5.3.2. Operation Costs

The cost of operation for hydropower stations consists of three main parts: Plant machinery maintenance including replacement of the part, insurance, and salary of staff. Hydropower plants have life ranging from 20 to 40 years and beyond. Typically a calculative lifetime of about 30 years is taken into account for financial calculations (Deshpande & Modak, 2002).

5.3.3. Maintenance Costs

There are two major parts of the maintenance cost of the machinery of the plant: breakdown maintenance and preventive maintenance. About 3–5% of the capital investment can be taken into account per year as maintenance and operation cost of station machinery in the primary years, which typically rises as the time passes.

Most of the maintenance work is required by the turbine runner part. Because of the hitting stroke of the silt incoming with the water or due to the cavitation effect, the runner blade becomes impaired. During the rainy season, this impairment is largest in some nations, as the holding time in the stilling basin or in the reservoir is abridged and when the arriving water in the reservoir carries a lot of eroded sand, the quality of water received at the turbine is comparatively meager. It is normal exercise during the period, for this reason where the demand of the power diminishes, to eliminate the runners of a multi-turbine power station for renovation (Gao et al., 2015; Laggoune et al., 2015).

Another significant cost element is insurance costs, in addition to maintenance. Insurance of dams and hydro turbines is needed to protect the damage of their huge capital partially or fully in conditions when components are impaired and also due to flooding, in conditions when parts are broken, and to shield the large potential loss if anything goes unplanned with the dam.

5.3.4. Plant Utilization Factor

In hypothetical calculations, whenever water is available, hydropower plants are considered to produce power. However, this may not be true in all circumstances in actual practices. The accessibility of hydropower stations has improved greatly during the last few decades with novel predictive and preventive measures of innovative control over plant. For about 95% of the time, the best plants may be accessible. For whole capacity usage of a hydropower plant, the unavailability of water though becomes a limitation. The plant consumption ranges from 60–80% due to this factor (Bobtana et al., 2019). In some viable locations, where the reservoir has a huge storage capacity, and when water accessibility is constant because of a mix of snow melting and rain-fed systems, a greater plant utilization factor may be attained. In the situation of small plants, on the other hand, this factor may even be lower than 50%.

5.3.5. Salary and Administrative Expenses

The factor of salary cannot be discounted in the financial examination, since a hydropower plant needs constant maintenance and monitoring. Old power plants used to need more control of numerous systems and persons to operate, whereas new plants need much less manpower due to automatic controls (Samimi et al., 2019).

5.4. ELECTRICAL TARIFFS

Different kinds of tariff systems are found to variations in policies in different locations worldwide, as defined below. Significant effects on the choice to familiarize a definite tariff system are open or non open markets, incentives for clean electricity generation, shortage or abundance of electricity available, etc (Muñoz et al., 2007; Caralis et al., 2012).

5.4.1. Feed-in Tariff

The scheme of feed-in tariff is founded on the rule of giving a quantity to the power generators, as its name suggests, depending on the extent of electricity which has been supplied to the grid. This mechanism works at already stated rate per unit of power. From a traditional (nuclear or fossil fuels) plant, this proportion is then the rate of generation of power. The grid controller cannot refute the reception of the electricity produced by the hydropower station although it is excess which is the supreme feature of the system of feed-in tariff (Kaldellis et al., 2005). Other non-hydro power stations, in this situation, should decrease their electricity production. The system of feed-in tariff is present, for example, in India, Germany and in various other states. Depending on the age and capacity of the power station, each controller of a hydroelectric power plant got a quantity in the rang of 3.5 Euro cent/kWh and 12.7 Euro cent/kWh in Germany.

5.4.2. System of Availability Based Tariff (ABT)

ABT is a system for the reclamation of fixed charges from transmission licensee or a power station by the marketable ways of disincentives or incentives. ABT is based on performance and is a tariff for the amount of power by producers controlled and possessed by those who are stakeholders in supplying electricity in many states or by the central government. It represens a novel mechanism of dispatching and scheduling electricity, which needs both beneficiaries and generators to obligate to daily schedules. Also, this is a mechanism of penalties and rewards pursuing to impose dailu pre-declared timetables, nonetheless disparities are allowed if informed one hours in anticipation. It aids in exchange of energy and capacity and simplifies grid discipline and also helps the meritocracy based distribution of eletrcity (Ayoub & Yuji, 2012).

Normally, there are three parts in ABT: A fixed charge (FC) is allocated each month to the generator by every receiver for creating capacity accessible for usage. For each beneficiary, the fixed charge might not be the equal. Usually, it changes with the stake of a receiver in the whole electricity production capacity. This foxed charge also changes with the level of accessibility attained by any producer.

The energy charge allocated for each kWh of energy provided is the second part as per a already declared schedule of the amount taken out on daily basis.

A charge for unscheduled interchange is the third part of ABT for the consumption and supply of energy in difference from the already declared daily timetable. This charge changes contrariwise with the frequency of the system prevalent at the time of consumption/supply.

5.4.3. Bulk Electricity Tariff System

Bulk tariff (for the power-producing company which supplies power to more than one state or central generating company) defines the annual fixed charges (AFC) for very hydropower plant which is decided by the Central Electricity Regulatory Commission. AFC calculation has the following parts (Candas et al., 2019):

- Return on equity.
- Interest on working capital.
- Depreciation.
- Maintenance and Operation expenses.
- Interest on loan capital.

The recovery of AFC is done in the system of energy charges (50% of AFC) and capacity charges (50% of AFC).

5.4.4. Time-Dependent Rates

In certain locations, during the times of maximum load at full utilization on the electrical grid, state-held electricity firms are concerned about purchasing electricity, because they consider in this manner that they can protect utilizing the electricity from the least effective producing units to generate power at greater prices. In some regions, therefore, contingent on the time of day, power companies put on flexible electricity tariffs, in case they purchase electrical energy from independent power station generators. Usually, lower than the regular user price of power is received by the power plant owners, becuase that price typically comprises payment for the power indutry's maintenance and operation of the electrical grid (Joskow, 2014).

The time of production of power becomes insignificant in the case of areas when a feed-in tariff system is present as in Germany.

5.4.5. Renewable Energy Certificates or Quota System

Provision of 'quota' of renewable energy is another mechanism of electricity tariff. In this mechanism, a 'quota' e.g., 20% is given to every generator of electricity for the grid. Each industry has to incorporate a 20% share of total electricity generated of energy reaching from renewable sources such as wind, hydro and solar. Penalties ought to be imposed if this quota does not meet the requirements of each company. In Europe, an adaptation of this system was discussed by the European Union. Systems are being established to grant certificates if any company genrates more than the allocated quota e.g., 25% from renewable sources, called Renewable Energy Certificate. Other companies can also purchase this certificate with a lesser part of renewables compared to the quota of i.e., 15%. The second company with the acquisition of the certificates wiould be taken into account to be conforming with this quota. The mechanism is ame as the idea of 'emission trading' (Weißensteiner et al., 2011). In Europe, up till now, no such agreement had been extended for announcing such a quota mehcanism. There has been a discussion on the topic comparatively expensive renewable energy technologies i.e., solar photovoltaic systems will not be chosen at all with such a quota system. As a result, the technological development of these technologies might get clogged. In late 2010, the Renewable Energy Certificates (REC) were initiated in India. It is based upon the philosophy of the Clean Development Mechanism (CDM) or a quota system. Production of a least percentage of renewable energy has been given as a target to every state. In a situation a state is not capable enough to meet the goal, it can purchase Renewable Energy Certificates from other generators to fulfil its discrepancy of renewable energy share. This is a same mechanism that works in the stock exchange.

5.4.6. Investment Incentives/Production Tax Incentives

The production tax incentive is a system founded on generation, that aids renewable energies by method od expense immunities from taxes of electricity, for example, the tax on energy, applicable to all generators, for renewable energies (Poh & Kong, 2002). Therefore, this mechanism meets the expense of an evaded cost on the produgenerator side. Moreover, as the name implies, the investment incentive is a system is for decreasing the price for the venture in renewable energies so as to attracts more reasonable capital.

5.4.7. Environmental Credit and Clean Development Mechanism

The usage of renewable energy resources is attractive for various power companies and governments around the world. Hence, they present a definite environmental value to renewable energy arriving from wind, solar, and hydropower plants, in industrial or developed countries, for example, in the system of repayment of taxes of electricity, etc. Due to the obvious fact in developing countries like India, that each unit of electricity produced evades production of the equal quantity of electricity in fossil fuel-based power stations, an opportunity is offered the avoided environmental emissions of revenues extra earnings by the Clean Development Mechanism under the Kyoto Protocol (Chen et al., 2013). One carbon credit represents one ton of CO_2 gas avoided or saved and vended to developed naations at charges of almost 15 Euro as of the year 2010.

5.5. HYDROPOWER AND OTHER ELECTRICITY GENERATION TECHNOLOGIES

For the supply of electricity, it is not normal for a country to depend on hydro. Hence, it is interesting to investigate in what manner hydro can be used along with other various technologies. Thermal production is signified by gas-fired, nuclear, coal-fire capacities (Duke & Kammen, 1999; Fischer & Newell, 2008). There is a lot of attraction for merging hydropower and other renewable energy technologies which are discontinuous too in Europe. Solar power, and wind power, and hydropower plants with no reservoirs are the most significant intermittent technologies. It is assumed that each technology can be demonstrated as a combined sector to make the analysis simple by the use of the cost function which shows a distinctive worth order of separate producers. Identifying a nuclear sector (super index *N*) and traditional thermal sector (super index *C*) (such as coal and gas), and contingent upon the cost of the changing chief energy source, a sector cost function with production level as the argument is assumed $c^C(e_t^C)$ and $c^N(e_t^N)$. For all the period's primary energy prices are assumed to be fixed with no technological change. All the cost which are not dependent on the output is exempted; as these costs should not affect the present usage.

Because based on quality, various technologies would be managed in the similar manner, the intermittent technologies can be taken together, with the current generation given by

$$e_t^I = \beta_t \bar{e}^I, \beta_t \in [0, \beta^{\max}] \qquad (19)$$

Here $ß_t$ is the current rate of movement of the recurrent energy source (sunshine, water, and wind) and e^{-I} shows the capacity in power units (kW) measured as average factor in the period in consideration, changing into energy (kWh) from capacity for the said period. It is assumed that only production independent costs are included, excluding variable costs in the intermittent technologies (Goulder & Schneider, 1999; Tuladhar et al., 2009).

A bathtub is placed in Figure 5.5, in the center of the illustration designated by the lowest streak from A to D, representing the hydropower sources for the two periods, and by walls established from these points.

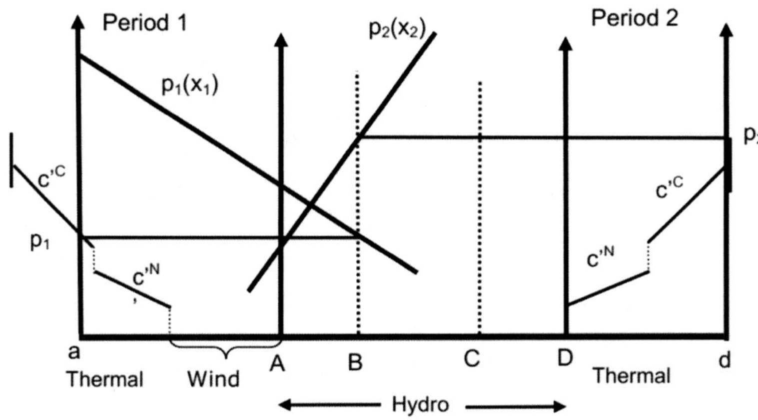

Figure 5.5. Comprehensive energy bathtub for thermal, intermittent, and hydro power.

[Source: https://www.researchgate.net/publication/5097288_Hydropower_Economics.]

Period 2 is calculated on the right-hand wall, and on the left-hand wall, period 1 is measured. The influx in the period 2 is denoted by CD, and the water resource accessible for period 1 is Ac which is made up of water inflow during period 1 and in the common case, water taken from the earlier thatne period 1. BC point gives the storage capacity for water and the reservoir capacity is shown by the walls established from these two points. The generation possibilities are stretched to the left of the wall of the bathtub for the period 1, specified for intermittent energy with marginal cost curves, succeeding the floor of protracted bathtub because the flexible price is zero, attached at the wall on the left-hand of the hydro bathtub, and the in worth order comes the peripheral price arc for traditional thermal capacities (c'^{C}) and e peripheral price arc for nuclear capacities (c'^{N}). Given capacity limits are indicated by the length of the curves. For simplicity, the cost curves are kept linear. Standard growing marginal price is exhibited by the marginal price curves. There is a leap to the cheapest traditional thermal capacity from the most affluent nuclear capacity (Wiser & Pickle, 1998). The similar image of the thermal marginal price arcs for period 1 is the addition of the hydro bathtub on the right-hand side for period 2, but by supposition, there is no alternating energy obtainable.

The consumption of electricity is calculated from right to left, and erected from point D, the demand arc for the period 2 is attached on the wall of energy at the right-hand (the attaching is not presented clearly). The electricity utilization is calculated from left to right, and erected from point A, the demand arc for period 1 is attached to the wall of energy on the left-hand. For ease of illustration, both demand arcs are shown linear. Period 2 is a high-demand period whearas period 1 is a low-demand period.

The settlement of the outside walls of protracted energy bathtub is *endogenously* measured implied by the optimum solution to the management issue. There the two walls are founded in such a way that steady illustrations are made with an ideal fundamental model solution of nature. The usage of all three technologies in period 1 is implied by the equilibrium price for period 1 (Johannsen, 2002). Traditional thermal technology is the peripheral technology in a way that whole capacity is moderately used and wind and nuclear are wholy used. A filled reservoir BC is left for period 2, and the quantity of water AB is the hydro contribution, locked in to be used in period 1. The price is the same as the peripheral cost of the moderately used traditional thermal capacity.

Without alternating energy in period 2, the condition is in a manner that thermal capacity is exploited completely. The influx in period 2, CD, and full water reservoir from period 1, BC. Hence, BD is used in period 2. The period of high-demand has a substantially greater value than the low-demand period, and due to full capacity utilization, the price is greater compared with the marginal cost of traditional thermal. Both thermal and hydro are utilized as peak load capacity. In the pure hydro case, a typical optimal solution would be the case that the value is equal in both the periods. This can still represent a usual condition, as presented in Figure 5.6. To study the outcome of changing wind source, on stream in period 1, no intermittent resource is assumed as shown in Figure 5.6. Enough production capacity is obtainable in period 2 as shown in illustrations to match values inside the capacity range of the traditional thermal capacity. Excess water, MC, is transported towards period 2 to profit from the wind resource and to maintain the same price. The same amount of the partly used technology will be used in each period which is an apparent outcome of equal price. Wind and water support the maximum load in period 2.

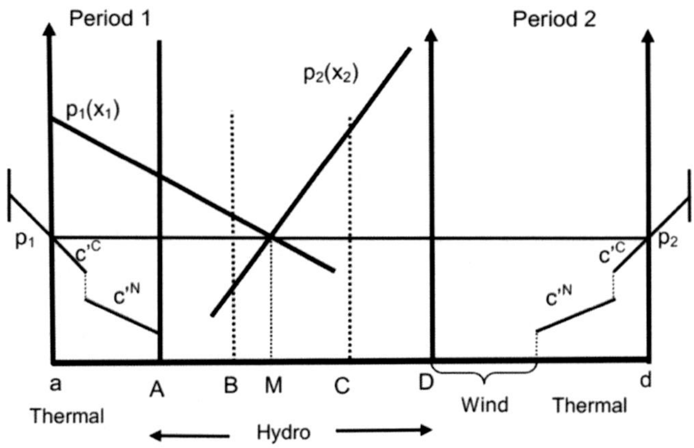

Figure 5.6. Discontinuous energy accessible just in period 2.

[Source: https://www.springer.com/gp/book/9781441944559.]

To study the outcomes of the introduction of a considerable amount of alternating power into a mechanism having abundance in hydropower, mainly wind power. The hydropower system of Norway is compared with an electric battery, in a Northern European context which could provide Europe in conditions when the wind is not gusting (Muizon & Glachant, 2004; Armaroli & Balzani, 2011). As seen from Figure 5.5, initiating trade between countries, Norway may trade hydroelectricity in times of high power demand in Europe (Germany and Denmark) and may also port the wind power utilized and save excellently by the usage of hydro during this period because of low wind. For Norway, it ought to make economic logic to export high and import less prices, but because without trade the condition would be much worse, this also profits the wind power nations.

It may be stated that in countries like Norway, the large-scale bringing of pumped storage in hydro plants facilitates a larger valuable line of work with wind-rich countries due to an increase in the inflows. The variation in electricity values with plentiful wind is the economic justification for pumped storage and it is insignificantly higher compared to the price of the loss in moving the water upr. The loss in energy ranges from 71 to 86%.

5.6. ADDITIONAL ISSUES IN HYDROPOWER

5.6.1. Uncertainty

Stochastic variables define the influxes to the reservoirs which is a very simple characteristic of the hydropower function. There are obvious problems created by this for hydropower management (102). A conclusion about the utilization of water, i.e., ferring water to the subsequent period, and production in the present period should be reached in the present period whereas the only the predictions can be used to find the future periods up to the possibility. Formulation of an optimal plan is the best that can be done in the current period by exploiting the *expectation* of the total of producer and consumer surpluses for every future period.

Early in the literature, the issue was recognized for finding the optimum solution to the hydro management issue posed by ambiguity. In Norway, a distinct solution method was presented by the Norwegian electricity regulator termed as the *expected water value approach* (Riddervold et al., 2020). One can discover several investigations on applied simulation and solution algorithms in the more specific engineering works. In the system-wide strategy for Norway, an approach to tackle uncertainty is used by large producers initiating in the approaches of the Norwegian electricity regulator.

Bringing in expectations for the next period is involved when making choices for the present period when influxes are recognized by taking uncertainty for the following period into account. Prices will change from their anticipations as time changes and this is a source of value deviation independent of limitations in the hydro system becoming active.

5.6.2. Market Power

The unguided electricity power generation system has motivated attentiveness in the potentials of generators working strategically in various nations from the early 1990s. Systems with an important share from hydro power having water storage have not been investigated mainly (Aasgård et al., 2016; Pinto et al., 2016).

Extremely economical immediate change in comparison with thermal energy producers make hydropower production suitable for strategic actions. Hydro producers interact daily in nations with day-ahead commercial markets and they all know that opportunity expense is signified by future predictable market values, functioning yield contingent on expenditures are zero, and they can grasp alike potentials.

5.6.3. Balancing Power

For the purpose of balancing of power, hydro producers make it a cost-efficient due to almost immediate opportunities of monitoring the generation. The essential time lag between the actual consumption and planning the production give rise to the need of balancing. Because of the effect of surrounding temperature, Demand can be based on random probability distribution, and on the manufacturer end, there can be events based on random probability distribution like mishaps of the transmission lines, mishaps inside generators or lines falling out or transformers, etc. malfunctioning. Availability to variable supply down or up is indispensable, as the supply and demand balance should be uninterrupted if short duration limit of demand has to be evaded (Brovold et al., 2004).

The pricing principles of hydropower is one of the issues serving balancing needs. Any variation in the generator's optimum schedule is imposed by applying the arbitrage principle up-regulation, and the social price is the change between highest price available in the future and the spot price at the moment of regulation, taking into account the capacity off storage of the reservoir. For down-guideline the social cost is zero provided that the producer has extra capacity for storage, but the social price is the spot price with a filled reservoir. It can be considered that the familiarization of more discontinuous energy will enhance the effectiveness of hydro in the matching market, but the case is not definite.

5.7. OUTLOOK FOR HYDROPOWER

The population of the world is expected to have an increase of about 50% by the year 2050, as associated to its value in 2000. The living standard of the inhabitants normally governs the energy intake per occupant per year. Today underdeveloped countries in the world, with about 2.3 billion occupants, have consumption of prime energy annual per capita of about 20 times lesser compared to developed countries which have about 1.4 billion occupants. Intake of electricity of underdeveloped nations per capita was about 35 times lesser compared to the usage in developed nations. Therefore, defenitly, the intake of the world especially the intake of electricity would probably enhance significantly in running century (Aasgård et al., 2016). Hence, the problem can be explained: the predictable rise in energy intake worldwide, a result of the combustion of fossil fuels, having the danger of a chief environmental effect, and global warming. A basic human right is a right for development, and without energy supply, there is no potential development. In this case, all accessible sources of energy will be required, the importance ought to be given the establishment of all the economically, and environmentally possible sources, including hydropower.

Apprehensions over uncertain pricing disrupting markets of fossil fuel, the important environmental outcomes of thermal energy sources, and the current decrease in nuclear energy considered a feasible energy source increased the importance of viable energy policies in last few decades whoch contain the important improvement of renewable energy provisions. From systems using solar energy, wind energy, geothermal energy, and comparable sources, electricity production is often related by common thinking to renewable energy. However, a time-tested and much proven technology offers the largest source of renewable energy: hydropower stations. Moreover, it also offers a possibility to accumulate energy with not transformation from its naturally accessible type (Mishkin & Adelman, 1994).

REFERENCES

1. Aasgård, E. K., Skjelbred, H. I., & Solbakk, F. (2016). Bidding methods for hydropower. *Energy Procedia*, *55*, 11–18.
2. Aasgård, E. K., Skjelbred, H. I., & Solbakk, F. (2016). Comparing bidding methods for hydropower. *Energy Procedia*, *87*, 58–78.
3. Abbasi, S. A., & Abbasi, N. (2000). The likely adverse environmental impacts of renewable energy sources. *Applied energy*, *65*(1–4), 121–144.
4. Abbasi, T., & Abbasi, S. A. (2010). Biomass energy and the environmental impacts associated with its production and utilization. *Renewable and Sustainable Energy Reviews*, *14*(3), 919–937.
5. Abrell, J., Kosch, M., & Rausch, S. (2019). Carbon abatement with renewables: Evaluating wind and solar subsidies in Germany and Spain. *Journal of Public Economics*, *169*, 172–202.
6. Ahmadzai, S., & McKinna, A. (2018). Afghanistan electrical energy and trans-boundary water systems analyses: Challenges and opportunities. *Energy Reports*, *4*, 435–469.
7. Alkherraz, A. M., Ali, A. K., & Elsherif, K. M. (2020). Equilibrium and thermodynamic studies of Pb (II), Zn (II), Cu (II) and Cd (II) adsorption onto mesembryanthemum activated carbon. *Journal of Medicinal and Chemical Sciences*, *3*(1), 1–10.
8. Armaroli, N., & Balzani, V. (2011). The hydrogen issue. *ChemSusChem*, *4*(1), 21–36.
9. Attia, E. A., Duquenne, P., & Le-Lann, J. M. (2014). Considering skills evolutions in multi-skilled workforce allocation with flexible working hours. *International Journal of Production Research*, *52*(15), 4548–4573.
10. Ayoub, N., & Yuji, N. (2012). Governmental intervention approaches to promote renewable energies—Special emphasis on Japanese feed-in tariff. *Energy Policy*, *43*, 191–201.
11. Azofra, D., Jiménez, E., Martínez, E., Blanco, J., & Saenz-Díez, J. C. (2014). Wind power merit-order and feed-in-tariffs effect: A variability analysis of the Spanish electricity market. *Energy Conversion and Management*, *83*, 19–27.
12. Bekchanov, M., Ringler, C., Bhaduri, A., & Jeuland, M. (2015). How would the Rogun Dam affect water and energy scarcity in Central Asia?. *Water International*, *40*(5–6), 856–876.

13. Benitez, L. E., Benitez, P. C., & Van Kooten, G. C. (2008). The economics of wind power with energy storage. *Energy Economics*, *30*(4), 1973–1989.
14. Bhattacharyya, S. C. (2010). Shaping a sustainable energy future for India: Management challenges. *Energy Policy*, *38*(8), 4173–4185.
15. Birol, E., Karousakis, K., & Koundouri, P. (2006). Using economic valuation techniques to inform water resources management: A survey and critical appraisal of available techniques and an application. *Science of the Total Environment*, *365*(1–3), 105–122.
16. Bjørner, T. B., & Jensen, H. H. (2002). Energy taxes, voluntary agreements and investment subsidies—a micro-panel analysis of the effect on Danish industrial companies' energy demand. *Resource and Energy Economics*, *24*(3), 229–249.
17. Boasson, E. L., Bohn, M., & Wettestad, J. (2009). CSR in the European oil sector: a mapping of company perceptions. *Corporate Social Responsibility in Europe: Rhetoric and Realities. Cheltenham, Edward Elgar*, 65–79.
18. Bobtana, F., Elabbar, F., & Bader, N. (2019). Evaluation of Halocnemum Strobilaceum and Hammada Scoparia Plants Performance for Contaminated Soil Phytoremediation. *Journal of Medicinal and Chemical Sciences*, *2*(4), 126–129.
19. Brovold, S. H., Skar, C., & Fosso, O. B. (2014). Implementing hydropower scheduling in a European expansion planning model. *Energy Procedia*, *58*, 117–122.
20. Brown, M. A., Zhou, S., & Ahmadi, M. (2018). Smart grid governance: An international review of evolving policy issues and innovations. *Wiley Interdisciplinary Reviews: Energy and Environment*, *7*(5), e290.
21. Candas, S., Siala, K., & Hamacher, T. (2019). Sociodynamic modeling of small-scale PV adoption and insights on future expansion without feed-in tariffs. *Energy Policy*, *125*, 521–536.
22. Caralis, G., Papantonis, D., & Zervos, A. (2012). The role of pumped storage systems towards the large scale wind integration in the Greek power supply system. *Renewable and Sustainable Energy Reviews*, *16*(5), 2558–2565.
23. Castelletti, A., Galelli, S., Restelli, M., & Soncini-Sessa, R. (2009). Tree-based reinforcement learning. *Water Resources Research*, *76*(7).

24. Castelletti, A., Galelli, S., Restelli, M., & Soncini-Sessa, R. (2010). Tree-based reinforcement learning for optimal water reservoir operation. *Water Resources Research*, *46*(9).
25. Chen, X., Wang, Z., He, S., & Li, F. (2013). Programme management of world bank financed small hydropower development in Zhejiang Province in China. *Renewable and Sustainable Energy Reviews*, *24*, 21–31.
26. Cocchi, G., Galli, L., Galvan, G., Sciandrone, M., Cantù, M., & Tomaselli, G. (2018). Machine learning methods for short-term bid forecasting in the renewable energy market: A case study in Italy. *Wind Energy*, *21*(5), 357–371.
27. Cramton, P., & Kerr, S. (2002). Tradeable carbon permit auctions: How and why to auction not grandfather. *Energy policy*, *30*(4), 333–345.
28. Cunha, J., & Ferreira, P. V. (2014). A risk analysis of small-hydro power (SHP) plants investments. *International Journal of Sustainable Energy Planning and Management*, *2*, 47–62.
29. da Silva, R. C., de Marchi Neto, I., & Seifert, S. S. (2016). Electricity supply security and the future role of renewable energy sources in Brazil. *Renewable and Sustainable Energy Reviews*, *59*, 328–341.
30. Darr, F. P. (1993). Critical Analysis of Joint Board Policy at the Federal Energy Regulatory Commission, A. *San Diego L. Rev.*, *30*, 485.
31. De Muizon, G., & Glachant, M. (2004). The UK Climate Change Levy Agreements: Combining negotiated agreements with tax and emission trading. *Voluntary Approaches to Climate Protection: An Economic Assessment of Private-Public Partnership, Edward Elgar, Cheltenham (UK)*, 231–248.
32. Deb, K. (1999). Multi-objective genetic algorithms: Problem difficulties and construction of test problems. *Evolutionary computation*, *7*(3), 205–230.
33. Deshpande, V. S., & Modak, J. P. (2002). Application of RCM to a medium scale industry. *Reliability Engineering & System Safety*, *77*(1), 31–43.
34. Dincer, I., & Acar, C. (2015). A review on clean energy solutions for better sustainability. *International Journal of Energy Research*, *39*(5), 585–606.
35. Dranka, G. G., & Ferreira, P. (2018). Planning for a renewable future in the Brazilian power system. *Energy*, *164*, 496–511.

36. Duke, R., & Kammen, D. M. (1999). The economics of energy market transformation programs. *The Energy Journal*, *20*(4).
37. Ekins, P., & Etheridge, B. (2006). The environmental and economic impacts of the UK climate change agreements. *Energy Policy*, *34*(15), 2071–2086.
38. Fischer, C., & Newell, R. G. (2008). Environmental and technology policies for climate mitigation. *Journal of environmental economics and management*, *55*(2), 142–162.
39. Førsund, F. R. (2005). Hydropower economics. *Memorandum*, *30*, 2005.
40. Gao, Y., Feng, Y., Zhang, Z., & Tan, J. (2015). An optimal dynamic interval preventive maintenance scheduling for series systems. *Reliability Engineering & System Safety*, *142*, 19–30.
41. Gatte, M. T., & Kadhim, R. A. (2012). Hydro power. *Energy Conservation*, *9*(51000), 95–124.
42. Glasson, J., & Salvador, N. N. B. (2000). EIA in Brazil: a procedures–practice gap. A comparative study with reference to the European Union, and especially the UK. *Environmental Impact Assessment Review*, *20*(2), 191–225.
43. Goulder, L. H., & Schneider, S. H. (1999). Induced technological change and the attractiveness of CO2 abatement policies. *Resource and energy economics*, *21*(3–4), 211–253.
44. Gyamfi, S., Derkyi, N. S., & Asuamah, E. Y. (2018). The Potential and the Economics of Hydropower Investment in West Africa. In *Sustainable Hydropower in West Africa* 328 (3), 95–107.
45. Hamelinck, C. N., Faaij, A. P., den Uil, H., & Boerrigter, H. (2004). Production of FT transportation fuels from biomass; technical options, process analysis and optimisation, and development potential. *Energy*, *29*(11), 1743–1771.
46. Harrison, G. P., Whittington, H. W., & Wallace, A. R. (2003). Climate change impacts on financial risk in hydropower projects. *IEEE Transactions on Power Systems*, *18*(4), 1324–1330.
47. Herbes, C., & Ramme, I. (2014). Online marketing of green electricity in Germany—A content analysis of providers' websites. *Energy Policy*, *66*, 257–266.
48. Hirsch, P. E., Schillinger, S., Weigt, H., & Burkhardt-Holm, P. (2014). A hydro-economic model for water level fluctuations: combining

limnology with economics for sustainable development of hydropower. *PloS one*, *9*(12), 114889.

49. Islam, M. R., Islam, M. R., & Beg, M. R. A. (2008). Renewable energy resources and technologies practice in Bangladesh. *Renewable and Sustainable Energy Reviews*, *12*(2), 299–343.
50. Johannsen, K. S. (2002). Combining voluntary agreements and taxes—an evaluation of the Danish agreement scheme on energy efficiency in industry. *Journal of Cleaner Production*, *10*(2), 129–141.
51. Joskow, P. L. (2014). Incentive regulation in theory and practice: electricity distribution and transmission networks. *Economic regulation and its reform: What have we learned?*, 291–344.
52. Kaldellis, J. K., Vlachou, D. S., & Korbakis, G. (2005). Techno-economic evaluation of small hydro power plants in Greece: a complete sensitivity analysis. *Energy Policy*, *33*(15), 1969–1985.
53. Kanellakis, M., Martinopoulos, G., & Zachariadis, T. (2013). *Energy Policy*, *1*, 79–90.
54. Kanellakis, M., Martinopoulos, G., & Zachariadis, T. (2013). European energy policy—A review. *Energy Policy*, *62*, 1020–1030.
55. Kelly, S. G., Vouras, M. F., & Amerkhail, J. S. (2005). The Subdelegation Doctrine and the Application of Reference Prices in Mitigating Market Power. *Energy LJ*, *26*, 297.
56. Kelman, M. P. N. C. R., Pereira, M., & Campodónico, N. (1998). Long-term hydro scheduling based on stochastic models. *EPSOM*, *98*, 23–25.
57. Kodikara, P. N., Perera, B. J. C., & Kularathna, M. D. U. P. (2010). Stakeholder preference elicitation and modelling in multi-criteria decision analysis–A case study on urban water supply. *European Journal of Operational Research*, *206*(1), 209–220.
58. Koirala, B. P., Koliou, E., Friege, J., Hakvoort, R. A., & Herder, P. M. (2016). Energetic communities for community energy: A review of key issues and trends shaping integrated community energy systems. *Renewable and Sustainable Energy Reviews*, *56*, 722–744.
59. Kong, L. B., Li, T., Hng, H. H., Boey, F., Zhang, T., & Li, S. (2014). Waste energy harvesting. *Lecture Notes in Energy*, *24*, 263–403.
60. Krarup, S., & Ramesohl, S. (2002). Voluntary agreements on energy efficiency in industry—not a golden key, but another contribution to

improve climate policy mixes. *Journal of Cleaner Production*, *10*(2), 109–120.

61. Laggoune, R., Chateauneuf, A., & Aissani, D. (2009). Opportunistic policy for optimal preventive maintenance of a multi-component system in continuous operating units. *Computers & Chemical Engineering*, *33*(9), 1499–1510.
62. Laughton, D. G., Sagi, J. S., & Samis, M. R. (2000). Modern asset pricing and project evaluation in the energy industry. *Journal of Energy Literature*, *6*, 3–46.
63. Lenzen, M. (2010). Current state of development of electricity-generating technologies: A literature review. *Energies*, *3*(3), 462–591.
64. Madani, K., & Mariño, M. A. (2009). System dynamics analysis for managing Iran's Zayandeh-Rud river basin. *Water resources management*, *23*(11), 2163–2187.
65. Madani, K., & Mariño, M. A. (2009). System dynamics for river basin. *Water resources management*, *4*(7), 145–157.
66. Mashayekhi, A., & Kazemi, A. (2019). Does the short-term boost of renewable energies guarantee their stable long-term growth? Assessment of the dynamics of feed-in tariff policy. *arXiv preprint arXiv:1907.11224*.
67. Matheussen, B. V., Granmo, O. C., & Sharma, J. (2019, July). Hydropower optimization using deep learning. In *International Conference on Industrial, Engineering and Other Applications of Applied Intelligent Systems* (pp. 110–122). Springer, Cham.
68. McCartney, M. P., Sullivan, C., Acreman, M. C., & McAllister, D. E. (2000). Ecosystem impacts of large dams. *Background paper, 2, 1–19.*
69. Menges, R. (2003). Supporting renewable energy on liberalised markets: green electricity between additionality and consumer sovereignty. *Energy Policy*, *31*(7), 583–596.
70. Ming, T., Liu, W., & Caillol, S. (2014). Fighting global warming by climate engineering: Is the Earth radiation management and the solar radiation management any option for fighting climate change?. *Renewable and Sustainable Energy Reviews*, *31*, 792–834.
71. Mishkin, M. T., & Adelman, D. I. (1994). Gas Industry Standards Board: Legal Considerations in the Standard Setting Process. *Energy LJ*, *15*, 73.

72. Mishra, S. K., Veselka, T. D., Prusevich, A. A., Grogan, D. S., Lammers, R. B., Rounce, D. R., ... & Christian, M. H. (2020). Differential impact of climate change on the hydropower economics of two river basins in high mountain Asia. *Frontiers in Environmental Science*, *8*, 26.
73. Mnushkina, O. I., Kas'yanov, A. A., & Samoilov, N. A. (2006). Optimization of the reforming process scheme. *Chemistry and technology of fuels and oils*, *42*(3), 176–182.
74. Mo, L., Lu, P., Wang, C., & Zhou, J. (2013). Short-term hydro generation scheduling of Three Gorges–Gezhouba cascaded hydropower plants using hybrid MACS-ADE approach. *Energy Conversion and Management*, *76*, 260–273.
75. Morillo, J. L., Zéphyr, L., Perez, J. F., Anderson, C. L., & Cadena, Á. (2020). Risk-averse stochastic dual dynamic programming approach for the operation of a hydro-dominated power system in the presence of wind uncertainty. *International Journal of Electrical Power & Energy Systems*, *115*, 105469.
76. Morillo, J. L., Zéphyr, L., Perez, J. F., Anderson, C. L., & Cadena, Á. (2020). Risk-averse stochastic dual dynamic programming approach for the operation of a hydro-dominated power system in the presence of wind uncertainty. *International Journal of Electrical Power & Energy Systems*, *117*, 105872.
77. Mosonyi, E. (1988). Water Power Development: Low-Head Hydropower Utilization. *Developments in Hydraulic Engineering*, 13 (4), 1–5.
78. Mousavian, H. M., Shakouri, G. H., Mashayekhi, A. N., & Kazemi, A. (2020). Does the short-term boost of renewable energies guarantee their stable long-term growth? Assessment of the dynamics of feed-in tariff policy. *Renewable Energy*, *159*, 1252–1268.
79. Mulat, A. G., & Moges, S. A. (2014). Assessment of the impact of the Grand Ethiopian Renaissance Dam on the performance of the High Aswan Dam. *Journal of Water Resource and Protection*, *20 (3), 1–14.*
80. Muñoz, M., Oschmann, V., & Tàbara, J. D. (2007). Harmonization of renewable electricity feed-in laws in the European Union. *Energy policy*, *35*(5), 3104–3114.
81. Niu, W. J., Feng, Z. K., Cheng, C. T., & Wu, X. Y. (2018). A parallel multi-objective particle swarm optimization for cascade hydropower reservoir operation in southwest China. *Applied Soft Computing*, *16*, 1–16.

82. Niu, W. J., Feng, Z. K., Cheng, C. T., & Wu, X. Y. (2018). A parallel multi-objective particle swarm optimization for cascade hydropower reservoir operation in southwest China. *Applied Soft Computing, 70*, 562–575.
83. Novak, W. J. (2019). Institutional Economics and the Progressive Movement for the Social Control of American Business. *Business History Review*, *93*(4), 665–696.
84. Omer, A. M. (2008). Ground-source heat pumps systems and applications. *Renewable and sustainable energy reviews*, *12*(2), 344–371.
85. Opgrand, J., Preckel, P. V., Sparrow, F. T., Thomas, G., & Loucks, D. P. (2020). Restoring the natural flow regime of a large hydroelectric complex: Costs and considerations. *Energy*, *190*, 116260.
86. Orhan, M. F., Dincer, I., Rosen, M. A., & Kanoglu, M. (2012). Integrated hydrogen production options based on renewable and nuclear energy sources. *Renewable and Sustainable Energy Reviews*, *16*(8), 6059–6082.
87. Pilavachi, P. A., Chatzipanagi, A. I., & Spyropoulou, A. I. (2009). Evaluation of hydrogen production methods using the analytic hierarchy process. *International Journal of hydrogen energy*, *34*(13), 5294–5303.
88. Piman, T., Cochrane, T. A., Arias, M. E., Green, A., & Dat, N. D. (2013). Assessment of flow changes from hydropower development and operations in Sekong, Sesan, and Srepok rivers of the Mekong basin. *Journal of Water Resources Planning and Management*, *139*(6), 723–732.
89. Pinto, T., Sousa, T. M., Morais, H., Praça, I., & Vale, Z. (2016). Metalearning to support competitive electricity market players' strategic bidding. *Electric Power Systems Research*, *135*, 27–34.
90. Poh, K. M., & Kong, H. W. (2002). Renewable energy in Malaysia: a policy analysis. *Energy for Sustainable Development*, *6*(3), 31–39.
91. Price, T., & Probert, D. (1997). Harnessing hydropower: A practical guide. *Applied energy*, *57*(2–3), 175–251.
92. Riddervold, H. O., Riemer-Sørensen, S., Szederjesi, P., & Korpås, M. (2020). A supervised learning approach for optimal selection of bidding strategies in reservoir hydro. *Electric Power Systems Research*, *187*, 106496.

93. Riddervold, H. O., Riemer-Sørensen, S., Szederjesi, P., & Korpås, M. (2020). A gradient boosting approach for optimal selection of bidding strategies in reservoir hydro. *arXiv preprint arXiv:2002.03941*.
94. Rietbergen, M. G., Farla, J. C., & Blok, K. (2002). Do agreements enhance energy efficiency improvement?: Analysing the actual outcome of long-term agreements on industrial energy efficiency improvement in The Netherlands. *Journal of Cleaner Production*, *10*(2), 153–163.
95. Ringkjøb, H. K., Haugan, P. M., & Solbrekke, I. M. (2018). A review of modelling tools for energy and electricity systems with large shares of variable renewables. *Renewable and Sustainable Energy Reviews*, *96*, 440–459.
96. Roggenkamp, M., & Boisseleau, F. (2005). The liberalisation of the EU electricity market and the role of power exchanges. *Regulation*, *4*, 2.
97. Roth, M., Eiter, S., Röhner, S., Kruse, A., Schmitz, S., Frantal, B., ... & Van der Horst, D. (2018). Renewable energy and landscape quality. *Pan*, *128*, 2–2.
98. Samimi, A., Zarinabadi, S., Shahbazi Kotanaei, A. H., Azimi, A., & Mirzaei, M. (2019). Use of data mining in the corrosion classification of pipelines in catalytic reforming units (CRU). *Quarterly Journal of Iranian Chemical Communication*, *7*, 681–691.
99. Sanchez, R. G., Seliger, R., Fahl, F., De Felice, L., Ouarda, T. B., & Farinosi, F. (2020). Freshwater use of the energy sector in Africa. *Applied Energy*, *270*, 115171.
100. Schaefli, B., Hingray, B., & Musy, A. (2007). Climate change and hydropower production in the Swiss Alps: quantification of potential impacts and related modelling uncertainties. *Hydrology and Earth System Sciences*, *11*(3), 1191–1205.
101. Schubert, M., & Boche, H. (2006). 5 Matrix-based SINR balancing and algorithmic solutions. *Foundations and Trends in Communications and Information Theory*, *2*(6), 482–482.
102. Singh, R. P., & Nachtnebel, H. P. (2016). Analytical hierarchy process (AHP) application for reinforcement of hydropower strategy in Nepal. *Renewable and Sustainable Energy Reviews*, *55*, 43–58.
103. Sodiq, A., Baloch, A. A., Khan, S. A., Sezer, N., Mahmoud, S., Jama, M., & Abdelaal, A. (2019). Towards modern sustainable cities: Review

of sustainability principles and trends. *Journal of Cleaner Production*, *227*, 972–1001.

104. Tarroja, B., AghaKouchak, A., & Samuelsen, S. (2016). Quantifying climate change impacts on hydropower generation and implications on electric grid greenhouse gas emissions and operation. *Energy*, *111*, 295–305.
105. Tuladhar, S. D., Yuan, M., Bernstein, P., Montgomery, W. D., & Smith, A. (2009). A top–down bottom–up modeling approach to climate change policy analysis. *Energy Economics*, *31*, S223-S234.
106. Wang, Q., Zhang, C., Ding, Y., Xydis, G., Wang, J., & Østergaard, J. (2015). Review of real-time electricity markets for integrating distributed energy resources and demand response. *Applied Energy*, *138*, 695–706.
107. Weißensteiner, L., Haas, R., & Auer, H. (2011). Offshore wind power grid connection—The impact of shallow versus super-shallow charging on the cost-effectiveness of public support. *Energy Policy*, *39*(8), 4631–4643.
108. Wiser, R. H., & Pickle, S. J. (1998). Financing investments in renewable energy: the impacts of policy design. *Renewable and Sustainable Energy Reviews*, *2*(4), 361–386.
109. Xue, B., & Sun, Z. (2018). Economics of hydropower energy: A critical assessment. *Energy Sources, Part B: Economics, Planning, and Policy*, *13*(1), 33–36.
110. Yang, M., Cela, B., & Yang, F. (2019). Innovative energy policy to transform energy systems in Ukraine. *Mitigation and Adaptation Strategies for Global Change*, 1–23.
111. Zakerzadeh, R., Bukac, M., & Zunino, P. (2016). Computational analysis of energy distribution of coupled blood flow and arterial deformation. *International Journal of Advances in Engineering Sciences and Applied Mathematics*, *8*(1), 70–85.
112. Zhou, F., Li, L., Zhang, K., Trajcevski, G., Yao, F., Huang, Y., ... & Liu, Q. (2020, August). Forecasting the Evolution of Hydropower Generation. In *Proceedings of the 26th ACM SIGKDD International Conference on Knowledge Discovery & Data Mining*, 238 (2), 2861–2870).

Chapter 6

Components of Hydropower Plants

CONTENTS

6.1. INTRODUCTION

There are numerous kinds of plants of hydropower categorized on diverse features. However, for each plant, few significant principal constituents are required and this chapter describes those constituents. The key elements of the hydro plant that are required to transform the water energy into electric power are explained here (Egré & Milewski, 2002; Sachdev et al., 2015). Lacking of these important parts makes the generation of electric power almost impossible. The main constituents of the hydroelectric power plant are given below.

- forebay;
- penstock;
- in-take structure;
- surge chamber;
- hydraulic turbine;
- upper and lower reservoir;
- powerhouse;
- tailrace; and
- draft tube.

6.1.1. Forebay

This is the basin region of any hydro plant where the water is stored momentarily before going into an in-take chamber. The water storage in a forebay is marked centered on the needed requirement of water in that particular region. This is utilized when the load necessity in the in-take is generally less (Singh & Singal, 2017).

It is known that water reservoirs are made across the river in order to stock the water, the stored water on an up-stream side of the dam can generally be carried out by the penstocks to a powerhouse. In this circumstance, the reservoir behaves as a forebay.

6.1.2. In-Take Structure

This is the structure that gathers the water running from the forebay and leads water into the penstocks. Diverse kinds of in-take structures are accessible and the choice of kind of in-take structure relies on several local circumstances (Wurbs, 1993).

In-take structure comprises some significant constituents of which the racks of trash play an important role. The racks of trash are delivered at the entry of penstocks in order to confine any fragments in the water.

If fragments flow with the water into a penstock then it will be the source of severe harm to the turbine runners, wicket gates, and turbine nozzles. The racks of trash are prepared from steel in the shape of rod. The rods are organized with the gap of 11 to 31 centimeters apart and the racks will isolate these fragments from the water which is running and whose permitted velocity is restricted to 0.7 to 1.6 m/sec (Gatte & Kadhim, 2012; Bergek et al., 2013).

In the regions of cold weather, there is a possibility of formation of ice in the water, to avert the entry of ice into the penstocks the racks of trash are heated with electric power and melts down the ice when it comes into contact with the racks of trash.

Other than the racks of trash, trolley and rakes organization which is utilized to wash these trash racks and the penstock shutting gates are given in the in-take structure.

Figure 6.1. Intake structure of the hydropower plant.

[Source: https://theconstructor.org/structures/hydropower-plant-components-functions/19705/]

6.1.3. Penstocks

These are just similar to large pipes placed having slope which transfers water from reservoir or in-take structures to the turbine. They operate with pressure so, abrupt opening or closing of the gates of penstock can trigger the water hammer impact to the penstock.

These are made to repel the effect of water hammer distant from the penstock is analogous to a usual pipe. In order to overpower the pressure, a huge wall is delivered for the penstock of short length and the surge tank is offered for the penstocks with long length (Gebre et al., 2013; Akella & Kumar, 2014).

Reinforced concrete or steel is utilized for building penstocks. For the small length, distinct penstock is utilized for every turbine likewise for the big length a large penstock is utilized and is divided into subdivisions at the end.

Figure 6.2. Penstock pipelines of the hydropower plant.

[Source: http://www.osakaeng.com/penstock/.]

6.1.4. Surge Chambers

A surge tank or chamber is the cylindrical tank that is generally open from the top in order to govern the pressure in the penstocks. It is joined to the penstock and closer to the PH (powerhouse).

Every time the PH disallowed the water load coming from the penstocks the water-level in a surge tank increases and governs the pressure in the penstocks (Angus,1937; Taylor & Strongman, 1985).

Likewise, when an enormous requirement is needed in the PH surge tank speeds up the flow of water into the PH and then the water level decreases. When the release is constant in a PH, the level of water in a surge chamber becomes persistent.

There exist diverse kinds of surge chambers accessible and they are chosen centered on necessity of the length of penstock and plant.

Figure 6.3. Surge chamber of the hydropower plant.

[Source: https://www.aboutcivil.org/surge-tanks-functions-types.html.]

6.1.5. Hydraulic Turbines

The hydraulic energy is converted into mechanical energy with the help of hydraulic turbine and then the mechanical energy can be converted into electrical energy by pairing the shafts of the turbine to a generator (Brandt, 1989).

The procedure adopted here is, when the water arriving from the penstock with high pressure strikes the runner or circular blades the water will move the shaft given at center and forces the generator to yield electricity.

Normally there are two types of hydraulic turbines namely:

- Impulse turbines; and
- Reaction turbines.

The impulse turbines are known as the velocity turbines. The Pelton wheel turbine is an instance of this turbine. The reaction turbines are known as the pressure turbines. Francis and Kaplan turbine are examples of this type.

Figure 6.4. A hydraulic turbine of the typical hydropower system.

[Source: https://www.alibaba.com/product-detail/Anti-cavitation-casted-casting-pelton-turbine_1363630809.html.]

The turbine can generally be well-thought-out as the main component of the plant of hydropower. Its part is to transform the power of water into mechanical power, by rotation of the shafts. The hydraulic turbine possesses a blades row fixed to the rotating plate or the rotating shaft. Running water, though passing over the hydraulic turbines, hits the turbine blades and the shaft is rotated because of the impact or variation of pressure and velocity (Silveyra et al., 2018). Whereas passing through the turbine the pressure and velocity of water lessen occasioning torque development and also rotation of the shaft of turbine. There are diverse designs or hydraulic turbines forms in utilization reliant on the functional necessities. The choice of a turbine is significant for the successful operation of the plant of hydropower (Gholikhani et al., 2019). The optimal output of the specific amalgamation of head and water flow can just be accomplished by a sufficient kind of hydraulic turbine.

6.1.6. Electricity Generator

Likewise, for a generator, the main purpose of the generator for the station of hydropower is to transform the rotation of shaft into electric power. After turbine, the significant part of the plant is an electric generator (Brownjohn et al., 2018).

Table 6.1. Synchronous Speeds of Generator in rpm

Number of poles	Number of pole pairs	Speed of Generator for 50 Hertz supply	Speed of Generator for 60 Hertz supply
60	30	100	120
20	10	300	360
18	9	333	400
14	7	428	514
2	1	3,000	3,600

The basic procedure of producing electric power in this way is to rotate the sequence of coils within the magnetic field or the other way round. The procedure triggers the electrons movement within conductors, which generates the electrical current.

The rotating speed of the synchronous generator that might be connected to the grid is kept persistent, and commanded by persistent frequency of the grid. The association amongst the frequency and rotating speed of the generator is administered by the given formula:

$$\text{Rotation} = \frac{\text{Frequency (50 Hz or 60 Hz)}}{2 \cdot \text{No. of poles}}$$

The amount of poles specifies the quantity of sets of the coils in a stator of a generator where the electricity is produced. 2 poles make 1 pair of pole therefore the poles quantity in the generator can't be the odd number. Because the point that the main grid is functioning with 3 phases, a generator should also produce 3 phase Ac. Just in the small plants that aren't linked to the main grid, the generators are of usually one phase, The dissimilarity is analogous to the point that majority of the large equipment or machines utilize 3 phase motors and the small equipment or machines utilize motors having one phase (Binama et al., 2017).

When utilizing this formula it should be considered that the frequency is measured in cycles per second and the rotation is measured in is rpm (revolutions per minute). Thus, the factor of conversion should be made practical in the equation above in order to transform frequency and rotation into the similar unit. Table 6.1 displays the variation needed generators speed with variation in several poles (Brownjohn et al., 2018).

The large hydro turbine speed isn't usually over 500 revolutions per minute, needing the generator of almost eight or more pairs of a pole.

The generator and turbine of the hydropower plant are joined through the shaft. Figure 6. 6 display the shafts of the 70 Megawatt machine. The turbine has a rotating speed of 500 revolutions per minute, and the generator has six pairs of pole for generating 50 Hertz current.

6.1.7. PowerHouse (PH)

A powerhouse in the building was given to shelter the electrical and hydraulic equipment. Usually, the entire equipment is sustained by a substructure or foundation placed for the PH.

In the situation of the reaction turbine, few machines such as scroll casing and draft tubes are fitted in the base whereas laying it. Therefore, the base is generally placed in very big dimensions.

When the turn of superstructure comes, the electric generators are installed on the mezzanine floor under which the vertical turbines are given. Apart from generators, horizontal turbines are also given. The control room is made on the mezzanine floor or the first floor.

Figure 6.5. Image of the power house at the hydropower plant.

[Source: https://theconstructor.org/structures/hydropower-plant-components-functions/19705/.]

The powerhouse and transformers of the plant of hydropower behave as the interface amongst the power transmission lines and electric generators

(Ossai et al., 2014). The voltage of produced middle voltage electric power by a generator, for example 6 thousand volts, is upsurged into the high voltage electric power by utilizing transformers. This kind of conversion is needed as higher voltages are favored for transmission of power over the long distances as the losses are decreased.

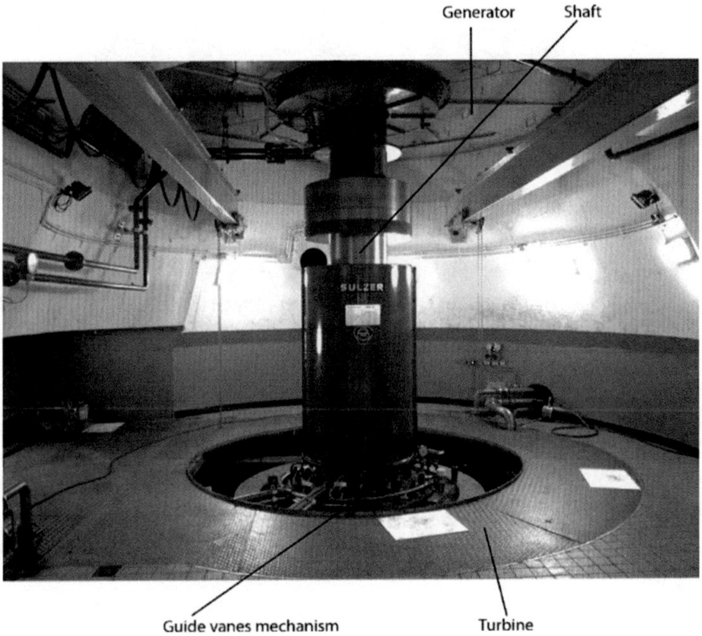

Figure 6.6. The shaft of seventy MegaWatt plant of hydropower situated in Roenkhausen, Germany.

[Source: https://link.springer.com/chapter/10.1007/978–3-642–20709–9_4.]

6.1.8. Draft Tubes

If the reaction turbine is utilized, then the draft tubes are the essential constituent that joins the outlet of the turbine to a tailrace. The draft tube comprises a steadily growing diameter in order to discharge the water into a tailrace with mild velocity. The opening gates are given at the end of the draft tubes which can generally be shut during maintenance.

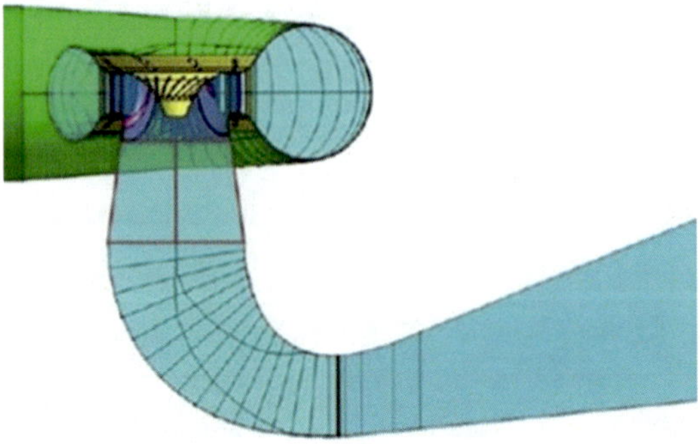

Figure 6.7. Schematic illustration of the draft tube.

[Source: https://theconstructor.org/structures/hydropower-plant-components-functions/19705/.]

6.1.9. Tailrace

It is the water flow from the turbine to stream. Tailrace is ideal if the PH is situated closer to the stream. If the powerhouse is situated far away from the stream then it becomes essential to construct the channel for transporting water into a stream.

Figure 6.8. Water flows from the turbines to stream through the tailrace.

[Source: https://energyeducation.ca/encyclopedia/Tail_race.]

Else the flow of water might harm the hydro plant in numerous manners such as lowering the efficiency of the turbine, damage to the blades of a turbine, cavitation, etc. (Ardizzon et al., 2014; Javed et al., 2020).

This is due to scouring or silting caused by needless water flow from the powerhouse. Therefore, the appropriate design of the tailrace must be more important.

6.1.10. Lower and Upper Reservoir

Storage kind of hydropower plants has the lower and upper level of water having the equipment in between. With respect to the word storage the water is kept in the reservoir before water reaches the turbines. The water is kept in an upper reservoir since throughout the course of day or with seasonal variations, diverse amounts of electric power are required from the main grid, and obtainability of the river water providing water to the plant might not resemble the condition of power controlling the water demand for the turbine. These kinds of reservoirs might be available naturally near the location of power generation; though, in many situations, they are only artificial reservoirs (Matos et al., 2019).

Figure 6.9. Sludge placed in the reservoir of one thousand MegaWatt storage power plant situated in Vianden, Luxemburg.

[Source: https://link.springer.com/book/10.1007/978–3-642–20709–9.]

Some of the storage power plants possess an additional reservoir of water at the release end of a turbine. This kind of release side reservoir is available in the power plants of pumped-storage. In this kind of power plant, the stored water in upper reservoir is transformed into electric power first when flowing down over the turbine. The water is kept in a lower reservoir and then pushed up to an upper reservoir in order to possess the potential energy when additional electric power is accessible in the non-peak times. As compared to an upper reservoir the lower reservoir is generally small and mainly depends upon factors like water availability and demand peaks of power in the profile of load, besides feasibility linked aspects like availability of the appropriate place for pinpointing the lower reservoir. Figure 6.9 displays the image of the emptied upper reservoir of 1 thousand MegaWatt power plant of pumped storage located in Luxemburg. The image displays the sludge quantity that gets accumulated at bottom of the reservoir of water which needs to be washed occasionally (Matos et al., 2019).

It may be observed that at times, because of the absence of the natural reservoir, a non-natural reservoir must be made.

6.2. STRUCTURAL COMPONENTS

These are components of the hydro station that don't take part directly in the generation of power; though, they make the fundamental structure that assists governed and safe water utilization for production of power by generator and turbine (Bueno & Carta, 2006; Vermaak et al., 2014).

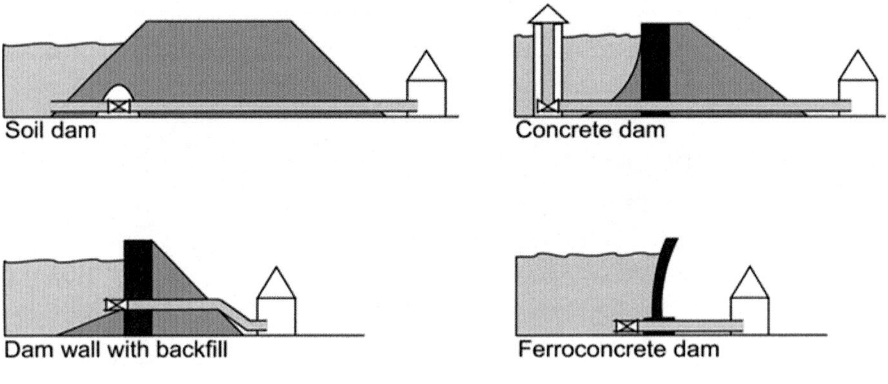

Figure 6.10. Four kinds of dam walls.

[Source: https://www.springer.com/gp/book/9783642207082.]

The parts that make up the major structure of any plant of hydropower are described below.

6.2.1. Dams and Spillways

Dams are made over the rivers in order to halt the flow of water and form the water reservoir. The water running down is stored in a reservoir as described in former section. The water is rerouted to turbine of the hydro stations. The dams gather water during the rainfall and stock it, therefore permitting for the constant flow of water through the turbines all over the year. The dams are utilized for governing floods and to preserve water for the purpose of irrigation. The main condition for the dams is to be capable to bear the applied pressure by the enormous amount of water kept in them. There are many diverse kinds of dams dependent upon the structure shape like arch dam, buttress dam, and gravity dam. The water height in a dam is known as the headrace (Abu-Sharkh et al., 2006).

For making the dams, the rivers must be rerouted from the construction sites and particularly should be evacuated so that the construction can happen actively. This is realized by the momentary reservoir bordered by the walls of metal. The area of the reservoir is generally kept arid and the work of construction is done quickly. As the dam structure spreads, at times it is needed to move the momentary reservoir concerning the plan of construction and the water inflow.

There are diverse kinds of the walls of dam, 4 of which are displayed in Figure 6.10. The soil dam is one of the first types. It comprises the trapezoid bank of stones and/or soil towards the lower and upper levels of water. The dam is roofed with the separation of flagstones of bitumen and/or clay.

One other kind is the dam of concrete. It is analogous to the previous kind but comprises the concrete pin and the concrete core which offers more firmness to the dam structure. The pin of concrete is located in the mid of a dam and ranges into the solid ground.

This type of dam was constructed first nearby 1850. The third dam kind is the dam having backfill. This kind doesn't have the form of trapezoid. The wall in front of the upper water is generally straight, whereas the backfill is generally on a lower waterside.

The 4th kind of dam wall displayed in Figure 6.10 is the ferroconcrete dam. This dam is the current constructional kind which is utilized for majority of the colossal dams, such as enormous dams for the storage plants. As the name specifies, the dam only comprises ferroconcrete and doesn't require additional backfill (Anderson et al., 2006; Sternberg, 2006). The precise shape of a wall is curved to a specific, computed degree of angle or bends which promises the maximum stability. Each dam possesses an opening pipe with the valve that might be placed at diverse places. The spillways are an essential part of the dam. The spillway is an approach for leaking water from the dams. It is utilized to offer for the discharge of floodwater from the dam. The spillway is utilized to avoid over filling of dams which might outcome in failure or damage to the dams. Spillways are usually of two kinds: uncontrolled and controlled kind. The uncontrolled kind of spillway begins discharging water when the water level rises above the certain level. In this kind, the water reaches the other end of a dam by overflow. In the situation of controlled kind, it is feasible to control the flow of water via gates given inside the structure of the dam that provides the outlet for discharging water to the other end of the dam without moving it over the turbines. Figure 6.12 displays the 60-meter high spillway of the hundred-year-old hydropower plant located in Schwarzenbach, Germany.

6.2.2. Surge Chamber

This is structure just like a tower that offer chambers for momentary water storage. The part of the surge chambers is to offer cushion space for supply or storage of water in circumstances of abrupt decrease or increase in the loading of turbine. This is the civil work before an inlet to the turbine that has controlling operations. When the valve of the hydro plant providing water to a turbine is unlocked, it might disturb the water head. In this kind of circumstance, the water pressure in the source pipe gets decreased and the likelihood of cavitation ascends (Pellegrino et al., 2020).

In contrast, when the valve of the hydropower plant is shut, because of an abrupt halt of the arriving water, the water pressure in the supply pipes increases abruptly and the supply pipes might burst. The chamber (Figure 6.13), which is a cylindrical water storing the configuration is joined to a pipe amongst turbine and dam and controls the swaying of water. When the valve of the hydro power plant is unlocked, this chamber can support the water it comprises to supply pipes to avert distribution of the water head. After the impact of an abrupt variation in water flow has calmed, the chamber is filled with water again to a similar level of an upper reservoir of water. When the plant's valve is closed again it controls the growing pressure by discharging water in the place above the former water level. Here the water's kinetic energy is transformed into the potential energy of water, which is the elevated water level. Once the water flow calms, the higher water level comes down automatically to the reservoir level (Poff & Hart, 2002; Azman et al., 2020).

Figure 6.11. Dam wall of the hoover station of hydropower, United States of America.

[Source: https://en.wikipedia.org/wiki/Hydroelectric_power_in_the_United_States.]

Figure 6.12. Dam wall along with spillway of hundred year old storing power plant situated in Black-Forest region, Germany.

[Source: https://link.springer.com/chapter/10.1007/978–3-642–20709–9_3.]

The mechanism and practice of function of the surge chamber of the plants of hydropower are primarily identified as recognized for the surge chamber utilized in the large water pump. The part of the surge chamber in every situation is to avoid damage of structure and equipment against the effects of variation in the loading of equipment.

6.2.3. Stilling Basin

The building of dams has a further significant structure that manages to overflow the stilling basin. The overflowing occurs from time to time and the plant of hydropower isn't aligned to every feasible level of water and thus the plant sometimes can't decrease the total quantity of water to a steady level. Therefore, as the portion of flood water can't cross the turbines, it should leave the reservoirs from someplace else. The role of the stilling basin is to some extent linked to the spillways.

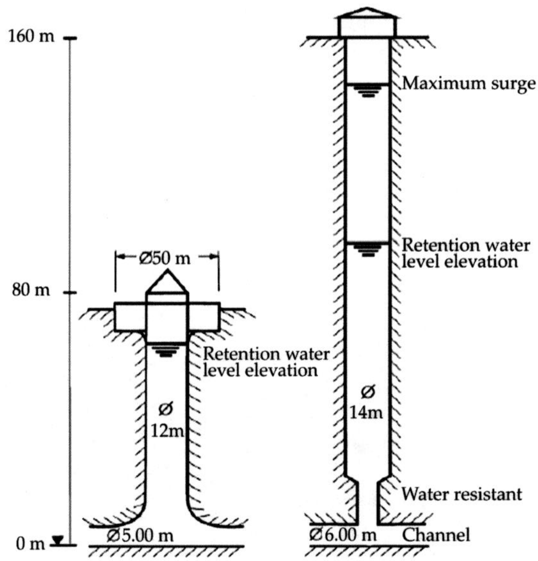

Figure 6.13. Surge chambers of the Schluchsee storing power plant situated in Germany.

[Source: https://link.springer.com/book/10.1007/978–3-642–20709–9.]

Whereas passing over a spillway, dependent upon the amount of flow, water can usually evoke huge electric power and it can normally be very tough to govern the destruction caused. The purpose of the stilling basin is to decrease that hazard (Chien, 1985). This is the kind of basin that possesses a particular form that guides the water into the anticipated intelligent manner and makes the water simpler to manage. There are several forms of the stilling basins; three kinds of the basin are defined here.

One likelihood to regulate the overflowing water is utilizing the steady basin with the pillar in mid (Figure 6.14). The flowing water must creek around this pillar and strikes together behind a pillar. Since there isn't any new method to make the passage, the water streams and falls over. Whereas this occurs the hazardous power of water is lost and can be brought carefully to a lower water level of a river (Yuan et al., 2012).

One more version of the stilling basins is by offering structures such as stars behind the dam. Whereas flowing down these stairs the water lessens its hazardous power before entering into the river again.

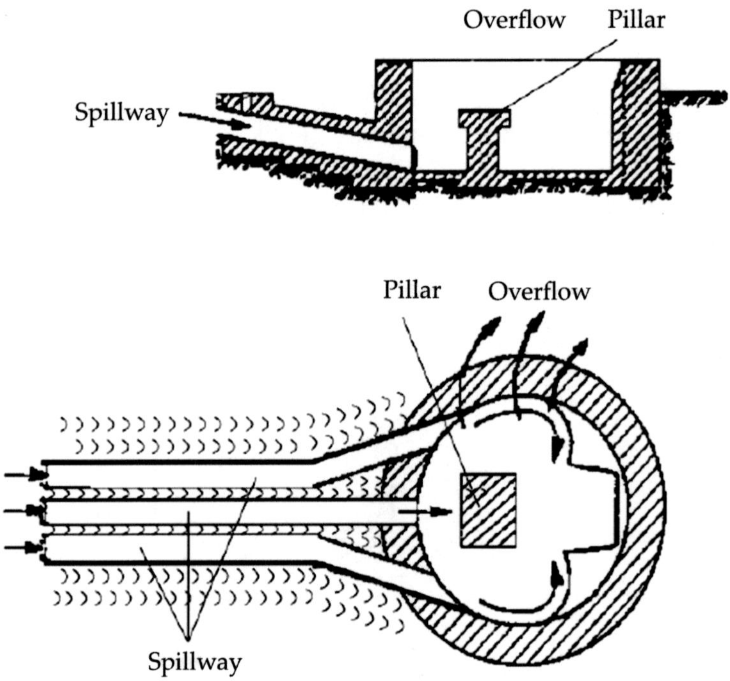

Figure 6.14. Stilling basin with a pillar.

[Source: https://link.springer.com/book/10.1007/978–3-642–20709–9.]

The third kind is the stilling basin having the slanting ground. When the water overflows the dam it flows into a basin and extends to every portion. This movement empowers the water to drop its power.

6.2.4. Spiral Casing and Penstock

Penstock is the pipe that transports water from a reservoir to turbine situated within the power station. Penstock is generally made of concrete or steel and is fortified with the systems of gate for regulating the flow. The water pressure through a penstock is quite high. In few locations where a hindrance is existent between the power station and dam like the mountain, the power station and the tunnel joining the reservoir serve the objective of the penstock (Dudgeon, 2000; Macklin et al., 2003). After flowing through the grill or screen for averting entry of the large solid objects like pebbles or stones, the water encounters an inlet shaped like trumpet known as spiral casing displayed in Figure 6.15. The cross-section of a spiral casing constantly reduces with the length circumferentially around a turbine. This construction part possesses the purpose to interrupt water on a turbine correspondingly, which confirms that each turbine part has a similar quantity of water. If a spiral casing wasn't present, the turbine blades that take water would get less amount of water as compared to the part attached to a turbine that comes before in the pathway of water. The inlet can normally be in a vertical or horizontal position and is made around Kaplan or Francis turbines. Moreover, they also can be prepared from concrete or steel or can also be self-supporting or made in bottom.

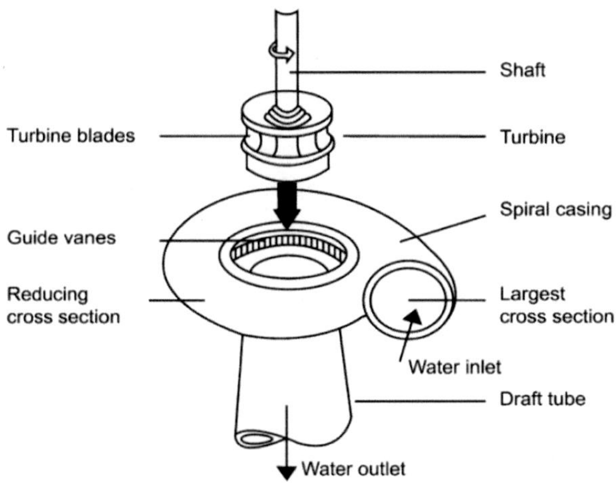

Figure 6.15. The spiral casing of Kaplan or Francis turbines.

[Source: https://theconstructor.org/practical-guide/francis-turbines-components-application/2900/.]

6.2.5. Tailrace

It is a downstream portion of the dam where the stored water enters into river again. It is the endmost part of a structure of a plant before the incoming water arrive the downstream lake or river. The tailrace extends towards the end for a purpose that with the expanding segment the losses of energy will be decreased than the losses in a persistent segment tailrace. The energy loss in leaving water aids to decrease the backpressure of water on an upstream side and subsequently aids in functioning the turbine more proficiently (Dudgeon, 2000; Macklin et al., 2003).

6.2.6. Pressure Tunnels or Pipes

Pressure tunnels or pipes, as the name recommends, guide water under the pressure and usually don't permit air to come into the system. The key objective of these pipes is to move water from a reservoir to the plant of hydropower. They are built by putting an extra layer of ferroconcrete amongst the deposits of rock and steel. The steel survives the pressure in a pipe, whereas the concrete joins the pipe to a rock. Under the pressure tunnel, detached from the real pipe, there exist extra grooves for evacuating the pipe nearby rock. Tunnels of free-level are less or more horizontal, whereas pressure tunnels can generally be vertical or sloping. The pressure tunnels might have diverse tilt angles varying from 31 to 91. What is more significant for pressure tunnels and pipes is a pressure head. Figure 6.16 displays the picture of 882-meter long pressure tunnels of the storage plant situated in the Black Forest region, Germany (Wilcox et al., 2014).

6.2.7. Cavern

Caverns, the structural portion of a hydro station, are built subterraneously inside the mountains. They can be made with the level under a lower level of water to avert cavitation, as described previously. There are various causes for building the plant inside the mountains. Firstly, there is a visual cause. The approval of the hydro station by neighboring inhabitants is even higher if the power plant is almost invisible. One more main reason is that the power plant situated inside the mountains is better secured. The third technical reason is that the downpipes are quite short (Haimes et al., 1998).

Cavern has additional tunnels that assist additional functions. Firstly, the drive tunnel is required to attain the diverse level in a power plant with consistent vehicles if the parts of machine require serving. An additional one is required for the ventilation of the power plant or the tunnel. The 3rd kind of tunnel guides the wires to their essential endpoints and conducts electric power. Therefore, there are 4 diverse kinds of tunnels: the ventilation tunnel, the drive tunnel, electricity tunnel, and the water-bearing pipe (Sertkaya et al., 2016; Morabito et al., 2020).

Figure 6.16. Pressure tunnels of the storing power plant situated in the region of Black Forest, Germany.

Figure 6.17 displays the hydro plant's cavern situated in Goldisthal, Germany. The utilization of caverns permits the plant of hydropower to be positioned under the mountain or hill. This kind of plant isn't visible from the outside.

Figure 6.17. Entry point of the drive tunnel of the caverns hydropower station from outside and inside.

[Source: https://www.visitscotland.com/see-do/unique-experiences-map/being-driven-into-a-mountain/.]

6.3. SUPPORTING PARTS

Furthermore to the key and the structural parts defined in the earlier two sections, there exist various parts that neither participate in the production of power nor establish the structural component of a power plant. Though, their utilization is significant for the control and operation of the plants. The constituents that are needed for the smooth and proficient operation of the hydropower plant are described below (Rinaldi et al., 2001; Zio, 2009).

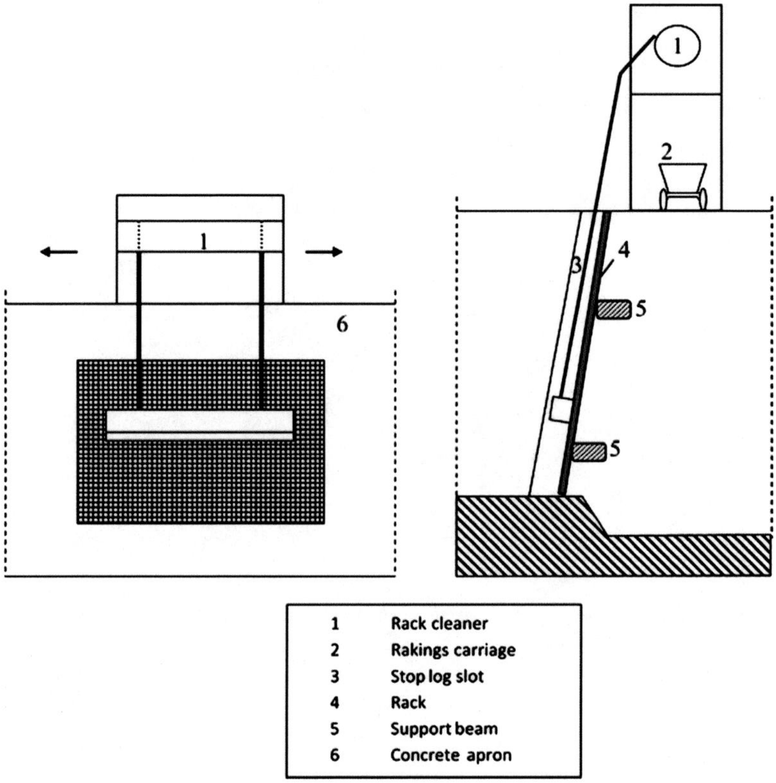

Figure 6.18. Grill cleaning machine.

[Source: https://link.springer.com/chapter/10.1007/978–3-642–20709–9_3.]

6.3.1. Screening Grills

The 1st device confronted by the stream of water moving towards a turbine is the grill (Figure 6.18). The grill has the purpose to secure water life or living species, which gives the meaning that any water animal or fish along with any solid like the ice or wood piece that are existent in water and are even bigger as compared to the holes in a grill will be averted from entering into the plant. If the screening grill isn't existing, the entrance of such kinds of objects into a turbine might cause harm to the blades. Though, this grill leads to specific energy loss via friction provided to the running water (Zhang et al., 2009). Whereas the grill gathers items that are bigger as compared to the gaps in a grill, like rubbish or branches, it also triggers extra resistance to the water flow which isn't favorable for the execution of a plant. To get free of the gathered objects, the plant has the grill washing machine. The crossbar shaped like shovel at bottom of a grill is dragged up and then fills then content into the wagon, which will be discarded later.

6.3.2. Control Gates

Control gates have a vital role by controlling the amount of the flow of water into a turbine via the penstock. The control gates are usually of vertical lifting kind and because of their large size and heavy weight, can be lifted only with the assistance of the large motors fixed on the top area of a dam. The main part of the control gates remains flooded in the water. As the control gate is in constant contact with corrosive circumstances, the material of control gates of the power plants is very significant and critical (Doolla & Bhatti, 2006). For a similar reason, the repair and maintenance of the control gate is the work-filled job associated with hydropower plants.

6.3.3. Control and Closing Valves

Shut-off valves are required in the hydroelectric plants for disturbing the flow of water during operation. This can generally be essential for safety problems regarding the bottom water, for evacuating the turbine if it needs maintenance or repair, along with in the power plants of pumped-storage when switching between pump working and turbine working.

There are three main kinds of shut-off valves described below and displayed in Figure 6.19. The 1st valve is a ball valve. This is analogous to the ball with the hole inside. When the water flows, the ball's hole is line up with the orientation of flow and the water rushes through a hole. If it is needed to stop the flow of water, the ball is then turned by 90 degrees (Adhikary et al., 2012).

Figure 6.20 exhibits an image of the control valve of the hydropower station. The connection from a penstock is visible on the left side and the water is entering a turbine spiral casing on the right side. The valve is heavy to survive the water pressure. It needs an extraordinary driving mechanism for the operation; lever and cylinders for regulating the position of a valve are also observed in the figure.

Another type of shut-off valve is a throttle valve. This is the disc that stays parallel to the orientation of flow when the flow of water is to be permitted. It is turned vertically by 90 degrees for stopping the water flow. To survive the force of water whereas stopping the water flow, this disc is shaped such as a lens, thicker in the center and thinner towards its edges.

The third type of shut-off valve is a turn valve which is analogous to the second one (throttle valve), the only dissimilarity being that this valve is turned from the other position.

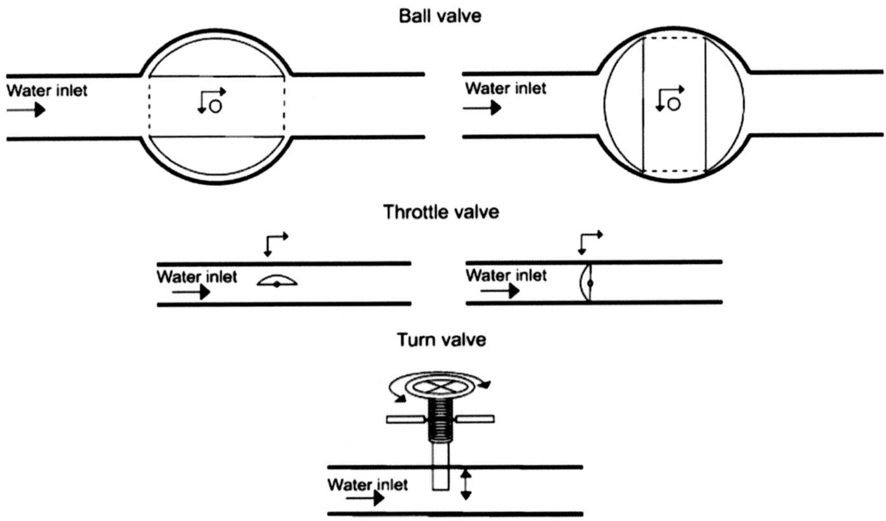

Figure 6.19. Types of the shut-off valves.

[Source https://link.springer.com/chapter/10.1007/978–3-642–20709–9_3.]

The functions of diverse valve types are indistinguishable; even though the respective constructions differ. In large plants, moving a ball, throttle valve, or valve is accomplished with the assistance of the electric motors and the pneumatic or hydraulic system. Turn valves aren't commonly utilized in daily operations (Mover et al., 1992).

6.3.4. Fish Passes

When the dam is built for the river power plant, humans interfere with the natural habitats of several species. The construction of the dam not only extends the river in order to take the shape of the water reservoir but stopping the water with the dam also averts water-living species from moving from one side to the other. To secure the life of water species, and to allow water species to pass through the river of power plants from upstream to downstream, fish passes are made (Hannett et al., 1999; Mansoor et al., 2000).

The concept of the fish passes is displayed in Figure 6.21. The fish pass offers structure just like stairs beside the dam. The fish passes are a series of very small water chambers having a deflector located interchangeably on the right or left side of the dam (Figure 6.22). The substitute deflectors aid to keep the water stream velocity comparatively low so that the fishes can survive whereas passing through them.

In the circumstance of more current power stations, small non-natural brooks running beside the dam wall are utilized as fish ways.

Figure 6.20. The control valve of a hydropower plant located at Roenkhausen, Germany.

[Source: https://www.springer.com/gp/book/9783642207082.]

The potential energy of the water passing through these fish passes isn't utilized for the generation of power, therefore it is well-thought-out as the loss of energy. Conversely, this is an essential contribution to the upkeep of the surrounding nature of the river power plant (Moran et al., 2018).

6.3.5. Guide Vanes

The ring of stationary modifiable vanes or blades (see Figure 6.23) constitutes a significant part of governing the operation of the hydro-turbine. The vanes are known as guide vanes as they escort the water in an appropriate direction concerning the blades of a turbine so that the flowing water enters at an anticipated angle.

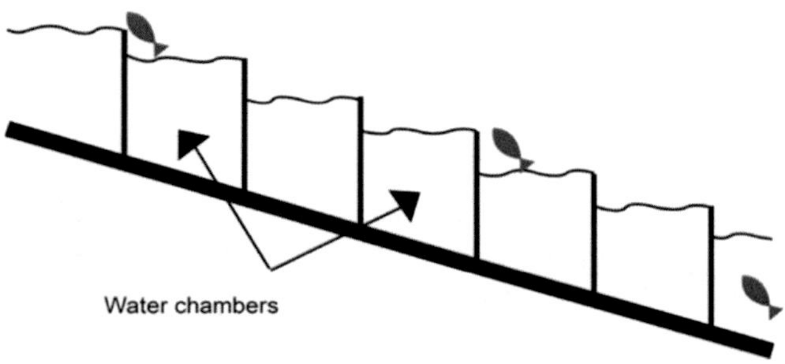

Figure 6.21. Scheme of the fish pass.

[Source: https://russianpatents.com/patent/237/2377365.html.]

Figure 6.22. Fish pass of the power station situated in Southern Germany.

[Source: https://www.springer.com/gp/book/9783642207082.]

Just after these stationary vanes, the water moves through the guide vane which can turn the water onto a turbine. The guide vanes can control the rate of flow by leaving space amongst each vane more closed or more open (Graf, 2001). The vanes are switched hydraulically with the assistance of a regulating ring which is moved in an anticlockwise direction and clockwise direction by the central impetus.

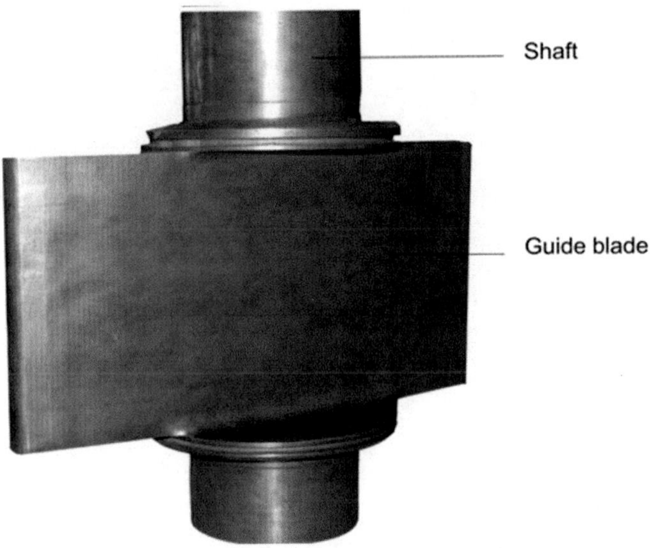

Figure 6.23. Adaptable guide vane with a shaft (hydropower plant Palmiet, South Africa).

[Source: https://www.mdpi.com/2227–9717/9/2/266/htm.]

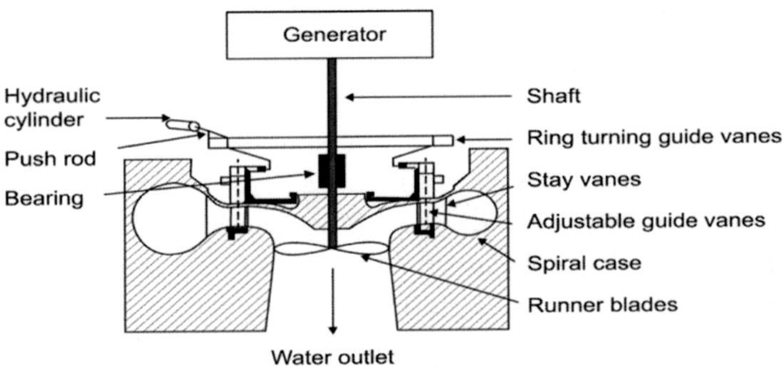

Figure 6.24. Scheme of the functioning mechanism of the guide vanes.

[Source: https://link.springer.com/chapter/10.1007/978–3-642–20709–9_3.]

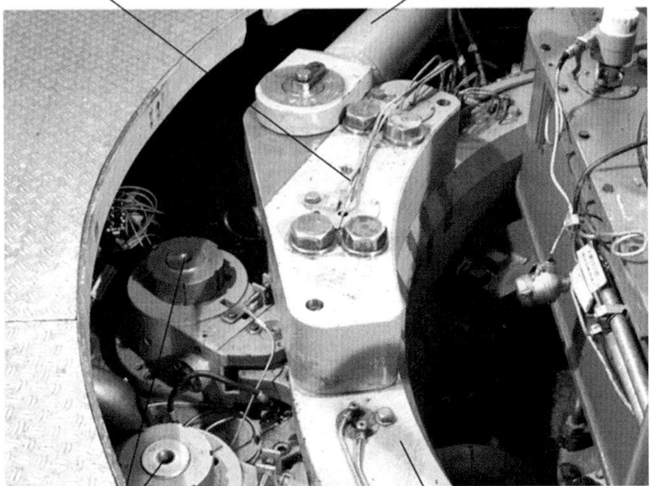

Figure 6.25. The functioning mechanism of the guide vanes of a power station situated at Roenkhausen Germany.

[Source: https://link.springer.com/chapter/10.1007/978–3-642–20709–9_3.]

After flowing through these guide vanes the water reaches the blades of a turbine, also known as runner vanes of rotor vanes (Hunt, 1979). Here the streaming water power drives a turbine by forcing the sloping shape of vanes. The precise rotation of the generator and turbine is governed by the shape of rotor blades along with regulating the water flow rate, which has been controlled by moving the guide vanes.

Figure 6.24 displays the schematic representation of the functioning mechanism of the guide vanes. For functioning these guide vanes, a big pushrod moves the rings, on which all of these vanes get cleaned by a similar angle around an individual shaft of each guide vane. It might be observed from the image that the mechanism needs the involvement of various subsystems like the lubrication system (Graf, 1999).

As stated at the start of the chapter, the hydropower plant might have thousands of small parts, though, the parts described in the chapter are the most significant ones, which are needed to understand the operation and control of the power stations (Graf, 1996; McDonald et al., 2004).

REFERENCE

1. Abu-Sharkh, S., Arnold, R. J., Kohler, J., Li, R., Markvart, T., Ross, J. N., ... & Yao, R. (2006). Can microgrids make a major contribution to UK energy supply?. *Renewable and Sustainable Energy Reviews*, *10*(2), 78–127.
2. Adhikary, P., Roy, P. K., & Mazumdar, A. (2012). Safe and efficient control of hydro power plant by fuzzy logic. *IJESAT*, *2*(5), 1270–1277.
3. Akella, A.K., & Kumar (2041). Report of the Committee of the Power Division on Ice as Affecting Power Plants. *Transactions of the American Society of Civil Engineers*, *95*(1), 1134–1150.
4. Allardice, N. W., & Thom, D. A. (1967). The design and construction of the Air New Zealand maintenance base at Auckland international airport. *New Zealand Engineering*, *22*(11), 435.
5. Anderson, E. P., Freeman, M. C., & Pringle, C. M. (2006). Ecological consequences of hydropower development in Central America: impacts of small dams and water diversion on neotropical stream fish assemblages. *River Research and Applications*, *22*(4), 397–411.
6. Angus, R. W. (1937). Water hammer in pipes, including those supplied by centrifugal pumps: graphical treatment. *Proceedings of the Institution of Mechanical Engineers*, *136*(1), 245–331.
7. Ardizzon, G., Cavazzini, G., & Pavesi, G. (2014). A new generation of small hydro and pumped-hydro power plants: Advances and future challenges. *Renewable and Sustainable Energy Reviews*, *31*, 746–761.
8. Azman, A., Ng, F. C., Zawawi, M. H., Abas, A., MAZ, M. R. R., Abustan, I., ... & Tam, W. L. (2020). Effect of Barrier Height on the Design of Stepped Spillway Using Smoothed Particle Hydrodynamics and Particle Image Velocimetry. *KSCE Journal of Civil Engineering*, *24*(2), 451–470.
9. Baxter, R. M. (1977). Environmental effects of dams and impoundments. *Annual review of ecology and systematics*, 255–283.
10. Bergek, A., Mignon, I., & Sundberg, G. (2013). Who invests in renewable electricity production? Empirical evidence and suggestions for further research. *Energy Policy*, *56*, 568–581.
11. Bergkamp, G., McCartney, M., Dugan, P., McNeely, J., & Acreman, M. (2000). Dams, ecosystem functions and environmental restoration. *Thematic Review II*, *1*, 1–187.

12. Binama, M., Su, W. T., Li, X. B., Li, F. C., Wei, X. Z., & An, S. (2017). Investigation on pump as turbine (PAT) technical aspects for micro hydropower schemes: A state-of-the-art review. *Renewable and Sustainable Energy Reviews*, *79*, 148–179.
13. Brandt, E. H. (1989). Levitation in physics. *Science*, *243*(4889), 349–355.
14. Brownjohn, J. M. W., Raby, A., Bassitt, J., Antonini, A., Hudson, E., & Dobson, P. (2018). Experimental modal analysis of British rock lighthouses. *Marine Structures*, *62*, 1–22.
15. Bueno, C., & Carta, J. A. (2006). Wind powered pumped hydro storage systems, a means of increasing the penetration of renewable energy in the Canary Islands. *Renewable and sustainable Energy Reviews*, *10*(4), 312–340.
16. Chien, N. (1985). Changes in river regime after the construction of upstream reservoirs. *Earth Surface Processes and Landforms*, *10*(2), 143–159.
17. Doolla, S., & Bhatti, T. S. (2006). Load frequency control of an isolated small-hydro power plant with reduced dump load. *IEEE Transactions on Power Systems*, *21*(4), 1912–1919.
18. Dudgeon, D. (2000). Large-scale hydrological changes in tropical Asia: prospects for riverine biodiversity: the construction of large dams will have an impact on the biodiversity of tropical Asian rivers and their associated wetlands. *BioScience*, *50*(9), 793–806.
19. Ebhota, W. S., & Inambao, F. L. (2019). Smart design and development of a small hydropower system and exploitation of locally sourced material for pelton turbine bucket production. *Iranian Journal of Science and Technology, Transactions of Mechanical Engineering*, *43*(2), 291–314.
20. Egré, D., & Milewski, J. C. (2002). The diversity of hydropower projects. *Energy Policy*, *30*(14), 1225–1230.
21. Gatte, M. T., & Kadhim, R. A. (2012). Hydro power. *Energy Conservation*, *9*(51000), 95–124.
22. Gebre, S., Alfredsen, K., Lia, L., Stickler, M., & Tesaker, E. (2013). Review of ice effects on hydropower systems. *Journal of Cold Regions Engineering*, 3, 196–222.
23. Gholikhani, M., Nasouri, R., Tahami, S. A., Legette, S., Dessouky, S., & Montoya, A. (2019). Harvesting kinetic energy from roadway

pavement through an electromagnetic speed bump. *Applied Energy*, *250*, 503–511.

24. Gibson, N. R. (1919). Pressures in Penstock Caused by the Gradual Closing of Turbine Gates. *Transactions of the American Society of Civil Engineers*, *83*(1), 707–740.
25. Graf, W. L. (1996). 18 Geomorphology and Policy for Restoration of Impounded American Rivers: What is' Natural?.' In *The Scientific Nature of Geomorphology: Proceedings of the 27th Binghamton Symposium in Geomorphology, Held 27–29 September, 1996* (Vol. 27, p. 443).
26. Graf, W. L. (1999). Dam nation: A geographic census of American dams and their large-scale hydrologic impacts. *Water resources research*, *35*(4), 1305–1311.
27. Graf, W. L. (2001). Dam age control: restoring the physical integrity of America's rivers. *Annals of the Association of American Geographers*, *91*(1), 1–27.
28. Graf, W. L. (2005). Geomorphology and American dams: the scientific, social, and economic context. *Geomorphology*, *71*(1–2), 3–26.
29. Haimes, Y. Y., Matalas, N. C., Lambert, J. H., Jackson, B. A., & Fellows, J. F. (1998). Reducing vulnerability of water supply systems to attack. *Journal of Infrastructure Systems*, *4*(4), 164–177.
30. Hannett, L. N., Feltes, J. W., Fardanesh, B., & Crean, W. (1999). Modeling and control tuning of a hydro station with units sharing a common penstock section. *IEEE Transactions on Power Systems*, *14*(4), 1407–1414.
31. Herbert, G. J., Iniyan, S., & Amutha, D. (2014). A review of technical issues on the development of wind farms. *Renewable and Sustainable Energy Reviews*, *32*, 619–641.
32. Howard, C. A. (1988). Specialist Services—Mechanical Ventilation, Air Conditioning, Lift Installations, Escalators and Fire Fighting Installations. In *An Introduction to Building Services* 128(2), 45–60.
33. Hunt, G. H. (1979). The evolution of fly-by-wire control techniques in the UK. *The Aeronautical Journal*, *83*(821), 165–174.
34. Javed, M. S., Ma, T., Jurasz, J., & Amin, M. Y. (2020). Solar and wind power generation systems with pumped hydro storage: Review and future perspectives. *Renewable Energy*, *148*, 176–192.

35. Kaunda, C. S., Kimambo, C. Z., & Nielsen, T. K. (2014). A technical discussion on microhydropower technology and its turbines. *Renewable and Sustainable Energy Reviews*, *35*, 445–459.

36. Kim, S. H. (2010). Design of surge tank for water supply systems using the impulse response method with the GA algorithm. *Journal of Mechanical Science and Technology*, *24*(2), 629–636.

37. Kondolf, G. M., Smeltzer, M. W., & Railsback, S. F. (2001). Design and performance of a channel reconstruction project in a coastal California gravel-bed stream. *Environmental Management*, *28*(6), 761–776.

38. LARINIER, M., & Travade, F. (2002). Downstream migration: problems and facilities. *Bulletin Français de la Pêche et de la Pisciculture*, (364), 181–207.

39. LAWRENCE, W. S., & Qamar, A. (1979). Hydraulic transients: A seismic source in volcanoes and glaciers. *Science*, *203*(4381), 654–656.

40. Likar, J., Žarn, J., Grøv, E., Čebašek, T. M., & Likar, A. (2015). temporary storage in big underground caverns. *RMZ–M&G*, *62*, 237–254.

41. Macklin, M. G., Brewer, P. A., Balteanu, D., Coulthard, T. J., Driga, B., Howard, A. J., & Zaharia, S. (2003). The long term fate and environmental significance of contaminant metals released by the January and March 2000 mining tailings dam failures in Maramureş County, upper Tisa Basin, Romania. *Applied Geochemistry*, *18*(2), 241–257.

42. Mansoor, S. P., Jones, D. I., Bradley, D. A., Aris, F. C., & Jones, G. R. (2000). Reproducing oscillatory behaviour of a hydroelectric power station by computer simulation. *Control Engineering Practice*, *8*(11), 1261–1272.

43. Markard, J., & Truffer, B. (2008). Technological innovation systems and the multi-level perspective: Towards an integrated framework. *Research Policy*, *37*(4), 596–615.

44. Matos, C. R., Carneiro, J. F., & Silva, P. P. (2019). Overview of large-scale underground energy storage technologies for integration of renewable energies and criteria for reservoir identification. *Journal of Energy Storage*, *21*, 241–258.

45. McDonald, A., Lane, S. N., Haycock, N. E., & Chalk, E. A. (2004). Rivers of dreams: on the gulf between theoretical and practical aspects

of an upland river restoration. *Transactions of the Institute of British Geographers*, *29*(3), 257–281.

46. Mohamad, H., Mokhlis, H., & Ping, H. W. (2011). A review on islanding operation and control for distribution network connected with small hydro power plant. *Renewable and Sustainable Energy Reviews*, *15*(8), 3952–3962.
47. Morabito, A., Spriet, J., Vagnoni, E., & Hendrick, P. (2020). Underground Pumped Storage Hydropower Case Studies in Belgium: Perspectives and Challenges. *Energies*, *13*(15), 4000.
48. Moran, E. F., Lopez, M. C., Moore, N., Müller, N., & Hyndman, D. W. (2018). Sustainable hydropower in the 21st century. *Proceedings of the National Academy of Sciences*, *115*(47), 11891–11898.
49. Mover, W. G. P., & Supply, E. (1992). Hydraulic turbine and turbine control models for system dynamic studies. *IEEE Transactions on Power Systems*, *7*(1), 167–179.
50. Ossai, C. I., Boswell, B., & Davies, I. J. (2014). Sustainable asset integrity management: Strategic imperatives for economic renewable energy generation. *Renewable Energy*, *67*, 143–152.
51. Pellegrino, R., Toledo, M. Á., & Aragoncillo, V. (2020). Discharge Flow Rate for the Initiation of Jet Flow in Sky-Jump Spillways. *Water*, *12*(6), 1814.
52. Poff, N. L., & Hart, D. D. (2002). How dams vary and why it matters for the emerging science of dam removal: an ecological classification of dams is needed to characterize how the tremendous variation in the size, operational mode, age, and number of dams in a river basin influences the potential for restoring regulated rivers via dam removal. *BioScience*, *52*(8), 659–668.
53. Price, T., & Probert, D. (1997). Harnessing hydropower: A practical guide. *Applied energy*, *57*(2–3), 175–251.
54. Ramezani, L., Karney, B., & Malekpour, A. (2016). Encouraging effective air management in water pipelines: A critical review. *Journal of Water Resources Planning and Management*, *142*(12), 04016055.
55. Riasi, A., & Tazraei, P. (2017). Numerical analysis of the hydraulic transient response in the presence of surge tanks and relief valves. *Renewable Energy*, *107*, 138–146.

56. Rinaldi, S. M., Peerenboom, J. P., & Kelly, T. K. (2001). Identifying, understanding, and analyzing critical infrastructure interdependencies. *IEEE Control Systems Magazine*, *21*(6), 11–25.

57. Rohlf, D. (2005). Global model approach for X-31 VECTOR system identification. *Journal of aircraft*, *42*(1), 54–62.

58. Sachdev, H. S., Akella, A. K., & Kumar, N. (2015). Analysis and evaluation of small hydropower plants: A bibliographical survey. *Renewable and Sustainable Energy Reviews*, *51*, 1013–1022.

59. Sertkaya, AA, Saraç, M., & OMAR, MA (2016). THE IMPORTANCE OF PUMPING STORAGE HYDROELECTRIC PLANTS FOR TURKEY. *Gazi Journal of Engineering Sciences (GMBD)*, *1* (3), 369-382.

60. Sherman, E. C., Torpen, B. E., Reed, O., Fox, H. K., Ewald, R. F., Scheidenhelm, F. W., ... & Stevens, J. C. (1929). Discussion of "Sherman on Experiments on Models of Pit River DAMS." *Transactions of the American Society of Civil Engineers*, *93*(1), 481–538.

61. Silveyra, J. M., Ferrara, E., Huber, D. L., & Monson, T. C. (2018). Soft magnetic materials for a sustainable and electrified world. *Science*, *362*(6413).

62. Singh, V. K., & Singal, S. K. (2017). Operation of hydro power plants-a review. *Renewable and Sustainable Energy Reviews*, *69*, 610–619.

63. Sternberg, R. (2006). Damming the river: a changing perspective on altering nature. *Renewable and Sustainable Energy Reviews*, *10*(3), 165–197.

64. Sun, Z., Huang, Q., Opp, C., Hennig, T., & Marold, U. (2012). Impacts and implications of major changes caused by the Three Gorges Dam in the middle reaches of the Yangtze River, China. *Water Resources Management*, *26*(12), 3367–3378.

65. Taylor, J., & Strongman, C. P. (1985). Section 2 Hydro-electric plant and equipment. *Small-Scale Hydro-Power: Watt Committee: report number 15*, 8.

66. Vermaak, H. J., Kusakana, K., & Koko, S. P. (2014). Status of micro-hydrokinetic river technology in rural applications: A review of literature. *Renewable and Sustainable Energy Reviews*, *29*, 625–633.

67. Wilcox, A. C., O'Connor, J. E., & Major, J. J. (2014). Rapid reservoir erosion, hyperconcentrated flow, and downstream deposition triggered

by breaching of 38 m tall Condit Dam, White Salmon River, Washington. *Journal of Geophysical Research: Earth Surface*, *119*(6), 1376–1394.

68. Wurbs, R. A. (1993). Reservoir-system simulation and optimization models. *Journal of water Resources Planning and Management*, *119*(4), 455–472.
69. Yuan, W., Yin, D., Finlayson, B., & Chen, Z. (2012). Assessing the potential for change in the middle Yangtze River channel following impoundment of the Three Gorges Dam. *Geomorphology*, *147*, 27–34.
70. Zhang, Z., Wu, Z., Wang, D., Xu, Y., Sun, Y., Li, F., & Dong, Y. (2009). Current status and technical description of Chinese 2× 250 MWth HTR-PM demonstration plant. *Nuclear Engineering and Design*, *239*(7), 1212–1219.
71. Zin, A. A. B. M., HA, M. P., Khairuddin, A. B., Jahanshaloo, L., & Shariati, O. (2013). An overview on doubly fed induction generators′ controls and contributions to wind based electricity generation. *Renewable and Sustainable Energy Reviews*, *27*, 692–708.
72. Zio, E. (2009). Reliability engineering: Old problems and new challenges. *Reliability Engineering & System Safety*, *94*(2), 125–141.

Chapter 7

Environment Impacts of Hydropower Plants

CONTENTS

7.1. INTRODUCTION

A hydropower plant is a source of renewable energy that produces electricity from water flowing down watercourses, as a consequence of rainfall. As part of a strategy to combat global warming, this makes it both sustainable and appealing (Warren, 2012; Mohamed, 2018). Likewise, hydropower and pumped storage hydropower can also be used to balance things other than renewable energy sources mostly on a grid, like solar and wind power. Therefore, hydroelectric generators, just like all the other power plants, have an impact on the environment.

Hydropower has a large impact, especially when a dam and reservoir are involved. Although the impact may be negative, this is not always the case. The advantages should easily outweigh the disadvantages of well-designed projects. Nonetheless, any project involving the level of stress that can be caused by the construction of a significant estuary scheme must be thoroughly evaluated and all repercussions considered before construction is approved (Warner et al., 2017).

The consequences of a dam and waterway project can be wide-ranging, as can the considerations. When a reservoir is built, what will be submerged? Will inhabitants be compelled to resettle? What impact will the dam have on the river's sediment flow? What effect will this have on downstream water flows? What are the consequences in terms of greenhouse gas emissions? Large strategies have the potential to impact entire regions, necessitating careful review at the regional and, in some cases, national levels. These impacts will, in certain situations, extend beyond national borders line (Kuipers et al., 2019).

However a run-of-river system is projected to cause less disruption, extensive environmental studies would still be required. Small hydropower techniques are rarely as troublesome as large hydropower projects, and their implications are usually limited to the immediate locality, allowing local rather than national choices to be taken. They should, however, be held to the same standards and criteria.

This issue is not new. For at least two millennia, humans have been altering watercourses, and some of the earliest structures are still standing. A rockfill structure from 1300 BC is still in service in Syria, while Roman dams may also be discovered in Spain. Dams were used to produce irrigation water, fresh water, and to assist manage watercourses in the past. Electricity generation has been only added to this list of uses since the late nineteenth century (Donohue, 2007; Wilkinson, 2014).

While dams will always alter the environment, the effects have historically been positive, resulting in an improvement in the living standards and conditions of those affected. Greater environmental awareness, combined with some reckless developments, resulted in a shift in perceptions toward the end of the twentieth century, and large hydropower projects have been thoroughly investigated much more closely and critically since then. The World Commission on Dams (WCD) was prompted to investigate what makes a good hydroelectric project as well as what makes a bad one.

Development and Dams: The WCD's efforts helped in the publication of A New Framework for Decision Making in 2000. Since then, there has been a reassessment of large hydropower, with greater acceptance of environmental concerns and widespread acceptance of many of the WCD's suggestions. At the same time, the linkage among the living standards and water projects of those harmed is becoming more universally recognized. This is particularly the case in Africa, whose hydropower growth is lower than that of other regions and standard of living are often poor. Following up on the WCD report, the International Hydropower Association released its Hydropower Sustainability Assessment Protocol in 2011, which included many of the WCD's recommendations (Morris et al., 1999; Tiwari, 2002).

Both documents acknowledge that a hydropower project, especially one of this size, will be disruptive, but that it does not have to be destructive. There will be significant environmental changes, people will be dislocated, as well as habitats will be demolished, but most of these ramifications can be managed in a sensitive manner. Displaced societies, for instance, are given a stake in the scheme and offered much better living standards, while swamped habitats are re-created besides the submerged area.

7.2. ENVIRONMENTAL IMPACT ASSESSMENT

Today, a comprehensive environmental impact assessment will almost always be required to make a case for a significant hydroelectric plant, and it will be obligatory in most instances. The project's effects, including any biodiversity effects, essential resettlement, seismic activity potential, and the implications on areas downstream of the project, will all need to be assessed (Vanclay, 2017). A study of this nature should also include recommendations for mitigating any adverse consequences of the development. In many cases, especially when international lending institutions are involved, a project will not be allowed to move forward unless the environmental assessment is appropriate. This holds for both private and public sector projects (Richter

& Thomas, 2007). Unfortunately, some projects are still not receiving the attention they deserve. There are a variety of reasons for this, including corruption, political gain or narrow self-interest, and national bodies' reluctance to accept international agencies' decisions. Furthermore, as the WCD admitted, there has been a potential to exaggerate the benefits and advantages of large-scale hydropower projects while downplaying the harm they cause.

7.3. SOCIAL IMPACTS: RESETTLEMENT

The need to relocate people whose residences or societies will be demolished will almost certainly be the most polarising effect of any hydroelectric power project. The Chinese authorities displaced 1.3 million people during the construction of the Three Gorges Dam, and more people may be forced to relocate if the reservoir banks become wobbly. This is one of the largest resettlement programs ever enacted for a hydroelectric plant, but even with many limited proportions, the impact on those engaged will be significant (Greve et al., 2015).

Just like in the situation of the three Gorges, resettlement statistics can be huge, but how can anyone assess if they were extremely excessive? Calculating the number of people displaced per megawatt of an existing power grid is one way of at least equating projects. The three Gorges had a 71:1 ratio. In Indonesia, the Kedung Ombo dam, a 29-MW project that displaced 29,000 people, had a 1000 people/MW ratio. In contrast, in the United States the Grand Coulee dam, constructed in the 1930s, had a ratio of 2.

Of course, it is critical to identify all those who will be affected. When it comes to people who live in the affected area, this may be forthright. However, many other people who scavenge in the area that will be flooded, for example, maybe difficult to identify. Civil works, in addition to the actual dam and watercourse, will affect the residents in a large project (Cernea, 2000). These, too, must be included.

If lots of people should be displaced, a fair thumb rule in recent developments is that they'll be financially better off than what they were previously. Those who have been resettled, moreover, should have a meaningful voice in where they are transferred as well as a share in the initiative. People will support a scheme if it can benefit a large number of people in society. If this is not the case, then development must be called into question. It is also crucial to handle relocation sensitively. In Laos, for example, residents of four villages were resettled in one village, but the new location suited two of the villages but not the other two due to their relative size and influence. Thus, when entire communities and their religious and cultural sites are at risk of being wrecked, the scenario becomes even more troublesome (Watson, 2009).

7.4. IMPACTS ON BIODIVERSITY

Even if a dam and basin do not relocate many people, if they affect a large area, they can have a huge impact on national biodiversity. In this situation, the comparative effect can be estimated by multiplying the flooded area by the generating capacity in megawatts. Three Gorges Dam used to have a 317 ha/MW ratio, however Grand Coulee Dam used to have a 5 ha/MW ratio. In the meantime, the Kompienga dam in Burkina Faso received the best score of any modern project, at 1426 ha/MW. Again, because the effect may vary based on the type of underwater area, this is merely a rough approximation. On the other hand large shallow basins, will be more influenced than deep, narrow basins by this measure (Rauff & Hatta, 2013).

The devastation of an species habitat by such a project is the single biggest risk to biodiversity. Even though hydro-projects take so long to build, it's necessary to develop a novel habitat to replace the one that is jeopardized while dam construction is underway, indicated that the threat is recognized. Plant species may find this relatively simple, but animal species may find it much more challenging. It is, however, possibleSeveral earlier projects are now including managed ecosystems that were not initially thought when the plants were built. Reservoirs and

Dams have a less obvious impact on aquatic animals, yet that can be just as destructive. Several rivers in France no longer support salmon as a result of dams, and the Yangtze dolphin was proclaimed vanished in China in 2006, partly due to hydroelectric advancements along the estuary, which prohibited the animal from moving.

Deoxygenation of deeper reservoir water has various consequences, such as affecting marine species that may normally live there. On the other hand, the creation of a basin can provide great possibilities for fish species, and large reservoirs can enable the development of fish farms, resulting in the creation of a previously undeveloped industry (Burroughs et al., 2006; Berman, 2013). When evaluating the impact on biodiversity, there is a risk of focusing on individual species while ignoring the cumulative impact that a scheme can have over a large area and a long period. The impact is sometimes unforeseeable and does not appear until years after the project is completed.

7.5. GEOLOGICAL IMPACTS

As a consequence of the pressure created at the surface, there is a mounting body of evidence that the building of dams and the reservoirs inundation can cause seismic activity in the strata (underlying). These kinds of effects are usually only seen with dams that are above 100 meters tall. The action is usually brief, although it might last for a long time in certain circumstances. A dam has been linked to a 2008 earthquake in China's Sichuan province. The earthquake led to the deaths of 80,000 individuals. The dam structure does not generate all of the pressures. The amount of water in the reservoir will create enormous subsurface pressure that did not exist previously (Moran et al., 2018).

Landslides in the reservoir's vicinity are also a threat. A similar incident occurred in Italy in 1983, when a reservoir overflowed, killing 2600 people. Such landslips have not only the ability to be catastrophic, as in this case, but they often reduce the volume of the reservoir, limiting its use. Furthermore, they damage habitat that was not previously impacted by the project.

7.6. HYDROLOGICAL IMPACTS

Every river has a load of tiny particles that are managed to pick up by the stream and carried downstream with it. Whenever a river is dammed, the silt often settles in the bottom of the reservoir's, eventually filling it up. The China's Sanmen Gorge hydroelectric plant reservoir on the Yellow River loses 40% of its size to sediment in about four years. While this is an extreme example, many reservoirs face this issue on some level.

Sedimentation will usually stabilise once enough material has been pushed much beyond dam to meet the quantity deposited annually. By occasionally opening the dam's sluice gates, it is sometimes feasible to wash sediment away. Sediment, on the other hand, is made up of abrasive particles, and its movement will cause wear when pass through the plant's turbines, necessitating replacement or repair (Silvius et al., 2000).

The reservoir effects are significant, but many of those significant effects on the environment of sediment deposition in a reservoir occur the dam downstream. The sediment loss on downstream habitats that rely on it is expected to be one of the most significant consequences. When the Aswan

Dam on the Nile was built, it prevented critical sediment from reaching the Nile delta. The depletion of such a material, which was the source of the delta's fertility, led in delta erosion as well as a rise in the demand for artificial fertilisers. The Nile delta's habitat is drastically different today than it was in the past. Identical outcomes can be found all across the world. For instance, issues in the Blac. Algal blooms in the Black Sea, for example, have been related to the depletion of sedimentary material caused by river dams that feed it, particularly the Danube (Dudgeon, 2000; Cernea & Schmidt, 2006).

Another result of the absence of sediment in the flowing water past the dam is that erosion at the dam site increases. The erosion downstream can be very devastating, and the damage will be especially obvious right after a dam is erected. With age, the riverbed may even be lined with rock(obdurate) as the downstream area is cleaned of all material that can be easily eroded. Rivers in other parts of the world frequently deepen and narrow as a consequence of erosion. Riverbanks can be truncated, causing damage to bridge foundations and other riverine buildings. A new steady state is attained several years following dam construction in the ideal case; however, this is not always the case.

A dam on a big river like the Yangtze or the Danube, for example, can help to manage floods throughout periods of heavy rain. Flood management is an important aspect of several dam projects, and it can keep downstream areas much safer. A dam also can enable the river upstream navigable where that was previously impossible, which can benefit local people (Reid et al., 2019).

7.7. GREENHOUSE GAS EMISSIONS

Hydropower projects are generally regarded as one of the most environmentally safe forms of energy generation. Greenhouse gas emissions were 1013 kg/MWh in amount, that is similar to emissions from wind power plants. Furthermore, not all hydroelectric projects are low polluters. Others can generate a lot of methane, which is a powerful greenhouse gas (Fernandes et al., 2016).

Methane is produced when organic matter accumulates in the bottom of a reservoir in which the water is de-oxygenated. Anaerobic digestion occurs under such conditions, producing methane gas. To avoid this, project designers should try to remove quite enough organic material from the area that will be flooded as feasible before submersion, by removing trees and removing vegetation where feasible. However, removing everything would be impossible. A Canadian utility, Hydro Quebec, that has looked into this, has established that production of methane from reservoirs follows a pattern. Production peaks between 3 and 5 years after the reservoir is occupied. Emissions are comparable to natural lakes after ten years. However, far higher quantities of methane release have been recorded in some circumstances.

The number of emissions generated may be determined by the kind of region where a reservoir is built. According to some recent studies in the United States, effluent from farming may be brought into reservoirs and feed to the bacteria that generate methane. The magnitude of such influences is currently uncertain (Gregory et al., 2006).

REFERENCES

1. Berman, S. (2013). Ideational Theorizing in the Social Sciences since "P olicy P aradigms, S ocial L earning, and the S tate." *Governance, 26*(2), 217–237.
2. Burroughs, H., Lovell, K., Morley, M., Baldwin, R., Burns, A., & Chew-Graham, C. (2006). 'Justifiable depression': how primary care professionals and patients view late-life depression? A qualitative study. *Family Practice, 23*(3), 369–377.
3. Cernea, M. M. (2000). Risks, safeguards and reconstruction: A model for population displacement and resettlement. *Economic and Political Weekly*, 3659–3678.
4. Cernea, M. M., & Schmidt-Soltau, K. (2006). Poverty risks and national parks: Policy issues in conservation and resettlement. *World Development, 34*(10), 1808–1830.
5. Chai, X., Tonjes, D. J., & Mahajan, D. (2016). Methane emissions as energy reservoir: context, scope, causes and mitigation strategies. *Progress in Energy and Combustion Science, 56*, 33–70.
6. Donohue, W. A. (2007). Methods, milestones, and models: State of the art in conflict analysis research. Negot. J., 23, 487.
7. Dudgeon, D. (2000). Large-scale hydrological changes in tropical Asia: prospects for riverine biodiversity: the construction of large dams will have an impact on the biodiversity of tropical Asian rivers and their associated wetlands. *BioScience, 50*(9), 793–806.
8. Fernandes, G. W., Goulart, F. F., Ranieri, B. D., Coelho, M. S., Dales, K., Boesche, N., ... & Soares-Filho, B. (2016). Deep into the mud: ecological and socio-economic impacts of the dam breach in Mariana, Brazil. *Natureza & Conservação, 14*(2), 35–45.
9. Freudenberger, H. J., & Robbins, A. (1979). The hazards of being a psychoanalyst. *Psychoanalytic Review, 66*(2), 275–296.
10. Graham, J. P., & Nachman, K. E. (2010). Managing waste from confined animal feeding operations in the United States: the need for sanitary reform. *Journal of Water and Health, 8*(4), 646–670.
11. Gregory, R., Ohlson, D., & Arvai, J. (2006). Deconstructing adaptive management: criteria for applications to environmental management. *Ecological Applications, 16*(6), 2411–2425.

12. Greve, H. R., Palmer, D., & Pozner, J. E. (2010). Organizations gone wild: The causes, processes, and consequences of organizational misconduct. *Academy of Management Annals, 4*(1), 53–107.
13. Gutiérrez, F., Parise, M., De Waele, J., & Jourde, H. (2014). A review on natural and human-induced geohazards and impacts in karst. *Earth-Science Reviews, 138*, 61–88.
14. Hill, R. P., & Stamey, M. (1990). The homeless in America: An examination of possessions and consumption behaviors. *Journal of Consumer Research, 17*(3), 303–321.
15. Ho, L., & Goethals, P. (2020). Research hotspots and current challenges of lakes and reservoirs: a bibliometric analysis. *Scientometrics, 124*(1), 603–631.
16. Holmén, H., & Jirström, M. (2009). Look Who's Talking! Second Thoughts about NGOs as Representing Civil Society. *Journal of Asian and African Studies,* 44(4), 429–448.
17. Hussain, A., Sarangi, G. K., Pandit, A., Ishaq, S., Mamnun, N., Ahmad, B., & Jamil, M. K. (2019). Hydropower development in the Hindu Kush Himalayan region: Issues, policies and opportunities. *Renewable and Sustainable Energy Reviews, 107*, 446–461.
18. Kuipers, K. J., May, R. F., Graae, B. J., & Verones, F. (2019). Reviewing the potential for including habitat fragmentation to improve life cycle impact assessments for land use impacts on biodiversity. The International Journal of Life Cycle Assessment, 24(12), 2206–2219.
19. Lash, G. G., & Blood, D. R. (2014). Organic matter accumulation, redox, and diagenetic history of the Marcellus Formation, southwestern Pennsylvania, Appalachian basin. *Marine and Petroleum Geology, 57*, 244–263.
20. Leira, M., Filippi, M. L., & Cantonati, M. (2015). Diatom community response to extreme water-level fluctuations in two Alpine lakes: a core case study. *Journal of Paleolimnology, 53*(3), 289–307.
21. Mohamed, N. N. (2018). Continuous Dispute Between Egypt and Ethiopia Concerning Nile Water and Mega Dams. In *Grand Ethiopian Renaissance Dam Versus Aswan High Dam 129*(2), 451–483.
22. Moran, E. F., Lopez, M. C., Moore, N., Müller, N., & Hyndman, D. W. (2018). Sustainable hydropower in the 21st century. *Proceedings of the National Academy of Sciences, 115*(47), 11891–11898.

23. Morris, N. G., & Parry-Jones, S. A. (1999). The affordability of water in an African town. *Water and Environment Journal, 13*(1), 1–6.
24. Platteau, J. P., & Abraham, A. (2002). Participatory development in the presence of endogenous community imperfections. *Journal of Development Studies, 39*(2), 104–136.
25. Rauff, S. A., & Hatta, Z. A. (2013). Armed Conflict Termination in Sri Lanka: An Opportunity to End Displaced Life and Renew Tamil-Muslim Relations. *Asian Social Science, 9*(2), 219.
26. Reid, A. J., Carlson, A. K., Creed, I. F., Eliason, E. J., Gell, P. A., Johnson, P. T., ... & Cooke, S. J. (2019). Emerging threats and persistent conservation challenges for freshwater biodiversity. *Biological Reviews, 94*(3), 849–873.
27. Richter, B. D., & Thomas, G. A. (2007). Restoring environmental flows by modifying dam operations. Ecology and society, 12(1).
28. Sharma, A. (2020). Mediation vis a vis Litigation: Is the time ripe to takeover or there is a need of balance of one on other. *PalArch's Journal of Archaeology of Egypt/Egyptology, 17*(7), 10780–10788.
29. Silvius, M. J., Oneka, M., & Verhagen, A. (2000). Wetlands: lifeline for people at the edge. Physics and Chemistry of the Earth, Part B: *Hydrology, Oceans and Atmosphere, 25*(7–8), 645–652.
30. Tiwari, D. N. (2002). Willingness to pay for improved water quality in Kathmandu. Valuing the environment in developing countries–case studies. Pearce, DW, Pearce, C. & Palmer, C.(eds). Edward Elgar: *Cheltenham, 98*(2), 130–160.
31. Tracy, E. F., Shvarts, E., Simonov, E., & Babenko, M. (2017). China's new Eurasian ambitions: the environmental risks of the Silk Road Economic Belt. *Eurasian Geography and Economics, 58*(1), 56–88.
32. Vanclay, F. (2017). Project-induced displacement and resettlement: from impoverishment risks to an opportunity for development?. *Impact Assessment and Project Appraisal, 35*(1), 3–21.
33. Warner, J. F., van Dijk, J. H., & Hidalgo, J. P. (2017). Old wine in new bottles: The adaptive capacity of the hydraulic mission in Ecuador. *Water Alternatives, 10*(2), 332–340.
34. Warren, G. S. (2012). Hydropower: It's a Small World After All. *Neb. L. Rev., 91*, 925.

35. Watson, V. (2009). 'The planned city sweeps the poor away…': Urban planning and 21st century urbanisation. *Progress in Planning, 72*(3), 151–193.
36. Whittington, D., Davis, J., & McClelland, E. (1998). Implementing a demand-driven approach to community water supply planning: A case study of Lugazi, Uganda. *Water International, 23*(3), 134–145.
37. Wilkinson, H. (2014). Sharing worlds: managing complex community relationships in challenging times. *Public Engagement and Social Science, 219*(6), 201–14.

Chapter 8

Hydraulic Turbines and the Use of Ocean Energies

CONTENTS

8.1. INTRODUCTION

The core of any hydropower plant is the turbine which transforms the water power into the shaft rotation by the generator which in turns produce electricity. The efficient transformation of the water power into rotation depends upon the suitable choice and working of the turbine. There can be various ways to classify the turbines of hydropower plants (Von & Brekken, 2019). Following are the three principal criteria for classification.

- Classification with respect to (w.r.t) direction of flow;
- Classification w.r.t pressure of water;
- Classification w.r.t shape and orientation of the turbine.

8.1.1. Classification with Respect to Direction of Flow

There are various flow paths from where the water can go over the hydraulic turbines. There are four main categories of hydraulic turbines with respect to the path of water flow that is as follows (Charlier, 2001; Khan et al., 2009):

1. Axial flow turbines

There is a parallel water flow path and an axis of rotation in this class of hydraulic turbines. The famous types of these turbines are Propellor and Kaplan turbines.

2. Radial flow turbines

The water flow of this class of hydraulic turbines is primarily in a plane which is at the right angle to the blades axis of rotation. Pelton turbine is an example of this class of turbine.

3. Mixed flow turbines

Practically, the direction of flow in most hydraulic turbines is neither completely radial nor completely axial. It holds a substantial element of both radial and axial flows. These kinds of hydraulic turbines are known as mixed flow turbines. The most famous type of mixed flow turbine is the Francis turbine where water invades in the radial path and moves out in the axial path (Mack et al., 1999).

4. Crossflow turbines

Water pass by the blade ring of cylindrical rotor (i.e., turbine wheel), in this class of hydraulic turbines, which seems similar to a blower wheel connected with an electric air heater. Hither, the twice energy is given by the water, to the lower and the blades of the upper turbine. An Ossberger turbine or a Banki turbine is the famous type of this turbine, which is named after their creators. The volume of water that is inserting into the turbine is controlled by a sliding valve. The usage of the crossflow turbines is just in the lower power range, which is less than 1 MW. The Ossberger-crossflow turbine is constructed being a multi-cell turbine in the situations where the water supply is required. In this case, the normal division is 1 to 2. The big and the small cell medium water flow is used by the small cell (Chen, 2010). This describes the reason for the realization of the high efficiency of the largely fluctuating water supplies.

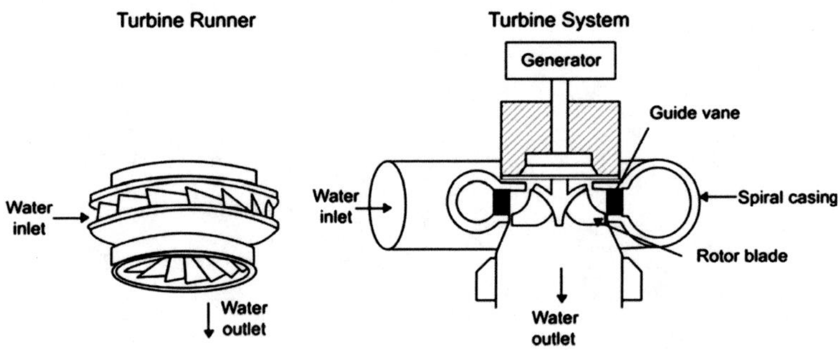

Figure 8.1. Diagrammatic illustration of a Francis turbine.

[Source: https://www.researchgate.net/figure/Sketch-of-the-Francis-turbine-model-and-the-LDA-setup-for-the-flow-survey-section-at_fig1_255470074.]

8.1.2. Classification w.r.t Pressure Change of Water

The hydraulic turbines can also be classified dependent on the variation in water pressure when it runs by the rotor connected to the hydraulic turbines. There are two types of hydraulic turbines dependent on the change of the pressure, (i) the impulse turbines, and (ii) the reaction turbines.

8.1.2.1. Impulse Turbines

The pressure of the water, in this type of turbines, does not change when running through the rotor of the turbine. But the change of the pressure is occurred, in impulse turbines, only in the system's nozzles that are never included in the rotor. Water is entered from a nozzle in an impulse turbine, therefore, a large fraction of its potential energy can be transformed into kinetic energy (Liu et al., 2015). After that, the high-speed jet has impinged on vanes of bucket-shaped attached to a rotating shaft, which converts the kinetic energy of the liquid into the rotary motion of the shaft. The Pelton turbine is the most famous kind of impulse turbine (refers to Figure 8.4) or popularly called the Pelton wheel (see Figure 8.5).

Figure 8.2. Deployment of a Francis turbine runner appearing in hydropower station ITAIPU.

[Source: https://www.wikiwand.com/en/Water_turbine.]

8.1.2.2. Reaction Turbines

The change of the water pressure, in this kind of turbines, is because of the variation in the flow path profile during its running along with the rotor blades. A reaction on the turbine blades is caused by the reduction in its pressure and the change in liquid's velocity, that is the working convention of such turbines. The liquid's reaction on the blades' rotation of the turbine is noticed; thus its name is derived as the reaction turbine. The most famous types of reaction turbines are Kaplan turbines and Francis turbines (Kumar & Saini, 2010; Date et al., 2013). The disparity within two types could be figured out by supposing the shaft impending a hinged door. The impulse turbines show a situation that when something hits the door with great velocity, and consequently, the door opens.

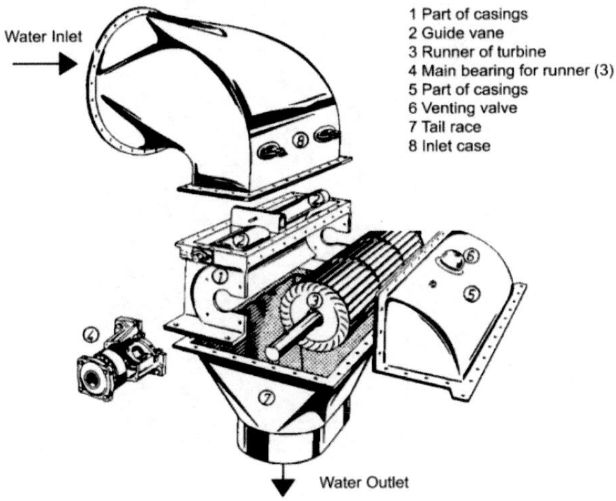

Figure 8.3. Illustration of a Crossflow turbine.

[Source: https://www.renewablesfirst.co.uk/hydropower/hydropower-learning-centre/crossflow-turbines/.]

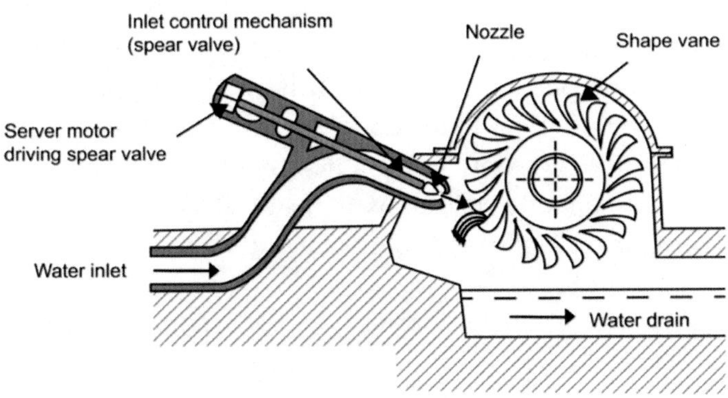

Figure 8.4. Illustration of a Pelton turbine.

[Source: https://www.renewablesfirst.co.uk/hydropower/hydropower-learning-centre/pelton-and-turgo-turbines/.]

While for the reaction turbine, the case is like forcing the door gently and adjusting the pushing force direction along with the altering direction of the door at each transitional step.

Figure 8.5. Pelton wheel (image captured by the reverence of Voith Hydro Holding GmbH & Co. KG).

[*Source:* https://tractionmech8.wordpress.com/2014/04/17/pelton-wheel-turbine/.]

For the first case, the door is opened faster if the velocity of the hitting object is greater, similarly, for impulse turbine, if the velocity of water is greater by which the bucket is hit, then it is rotated faster and, as a result, more power is provided. In the second case, a high force is required and requires to act in the appropriate direction same as for the case regarding reaction turbine. A separate section explains these two types in more details (Wagner & Mathur, 2011; Momčilović et al., 2012).

8.1.3. Classification with respect to Orientation and Shape of Turbines

Turbines can be classified by their manufacturing or deployment as well. There are vertical turbines, Straflo turbines, and bulb turbines.

Vertical turbines are about vertically oriented and the generator is placed above the water current (refers to Figure 8.7). Kaplan and Propeller turbines are considered to be examples of such types of turbines.

The bulb turbines, on the other hand, are about horizontally oriented and their generator is situated in a case shaped similar to a pear or a bulb. Hither the construction is surrounded by the streaming river water partially and then runs through the turbine (refers to Figure 8.6). Kaplan and Propeller turbines are concerned (Momčilović & Odanović, 2019).

An advanced bulb turbine is the Straflo turbine, with the addition of the generator poles on the motor's external ring. The rotor blades are hooked to a ring that triggers the generator. Propeller turbines are viewed to be Straflo turbines.

The Francis and Kaplan turbines can be either placed vertically or horizontally. Many big Francis turbines are viewed as rotating in a horizontal position because of the improved operation of spiral casing in a horizontal plane. Turbines, known as Pelton turbines, are utilized for high heights, however, water's low masses. Mostly, these are moving in a vertical position.

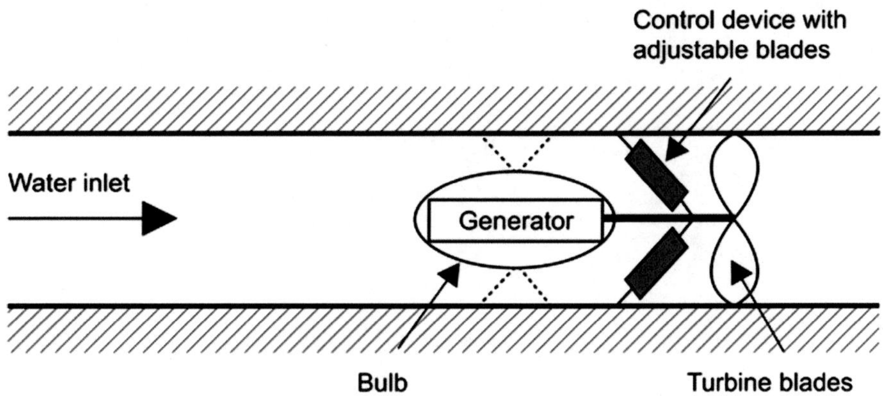

Figure 8.6. Illustration of a bulb turbine (horizontal turbine).

[Source: https://link.springer.com/chapter/10.1007/978–3-642–20709–9_5.]

Furthermore, to the techniques of the categorization of turbines that are discussed over, turbines can be categorized by the loading degree as well. They can be either fully loaded or partially. When the water runs to the entire rotor periphery then full loading occurs, whereas partial loading takes place when the water flows just on a few regions to the rotor periphery. Consequently, turbine blades having partial loading holds on a "rest" period within the loadings. The Pelton turbine is an example of this kind of turbine, where just a few blades are functional at either single unit of time (Anderson et al., 1993; Braitsch & Haas, 1996).

8.2. THEORY OF HYDRO-TURBINES

Moreover, the classification of turbines can also be based upon their kind of construction, the Francis, Kaplan, and Pelton, or Propeller turbines are the important ones which will be discussed further in the next chapters. Their names are after their inventors' names (Anilkumar et al., 2018).

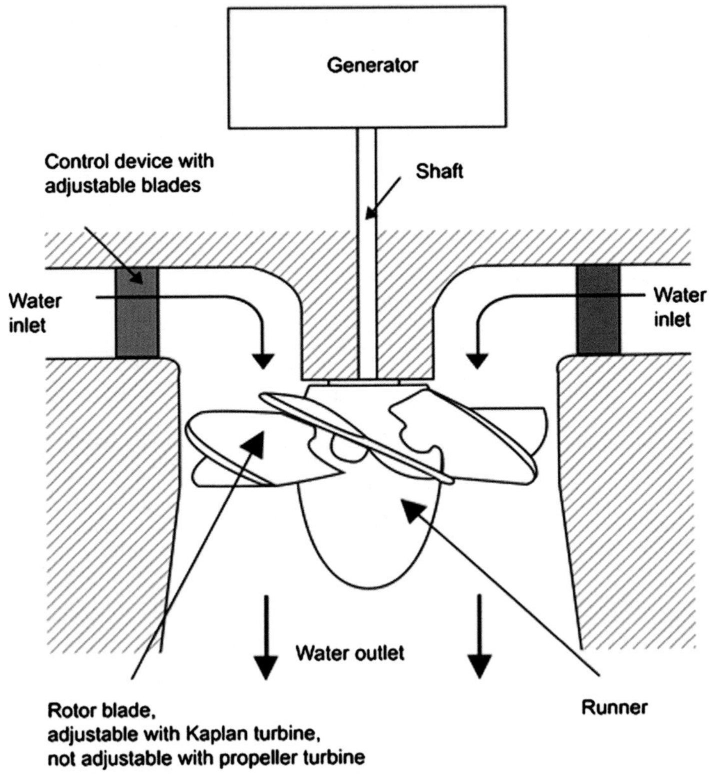

Figure 8.7. Diagrammatic view of a vertical turbine.

[Source: https://link.springer.com/chapter/10.1007/978–3-642–20709–9_5.]

8.2.1. Francis Turbines

In 1849, a scientist named James Francis has been developed the Francis turbine in England. It looks like a ring cake shape. Generally, the Francis turbines are utilized for mild heights of a drop from around 15.0 up to 500.0 m, also a water stream of up to around 500 m^3/s. Practically, the Francis turbines are usually applied for pumped storage power plants, hence, they can be utilized for pumping as well (Shrestha & Choi, 2020).

The Francis turbine is referred to as a reaction turbine. The energy is left when the water flows via the turbine. The momentum of the water flow is changed and its energy is given to the turbine blades surface that results in the pressure change because of their profile. The location of the turbine is within the low-pressure water get off, and the high-pressure water source.

Figure 8.8. Francis turbine runner (image captured by the reverence of Voith Hydro Holding GmbH & Co. KG).

[*Source:* http://me-resetsg.blogspot.com/2016/02/francis-turbine-runner.html.]

Figure 8.9. Creating of a runner of a Francis turbine within a Chinese factory.

[Source: https://www.springer.com/gp/book/9783642207082.]

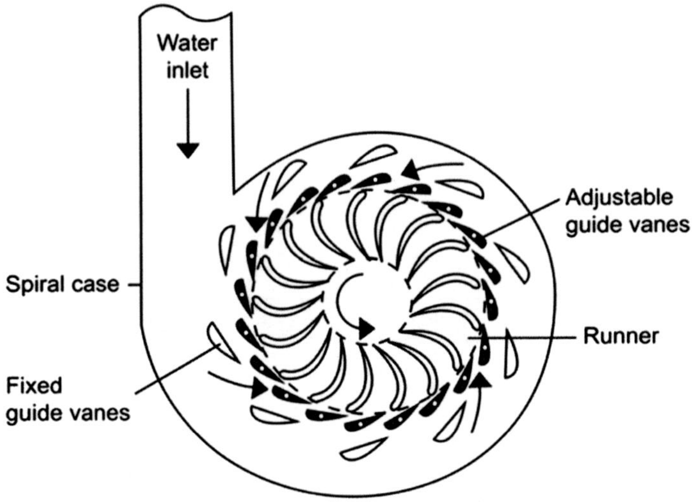

Figure 8.10. The structure of blades of a Francis turbine (top picture).

[Source: https://nptel.ac.in/content/storage2/courses/112104117/chapter_7/7_7.html.]

A spiral-shaped entry is possessed by the entry of water into the turbine. Within the spiral casing, the water leaves behind a part of its pressure which keeps its speed. The water is turned tangentially to the turbine wheel by the guide vanes that are located right after next to the runner (refers to Figure 8.10), out of significant effect. Such radial flow serves on the runner vanes, which force it to rotate. Water pressure reduction with the runner blades is the reason for the turbine rotation, which bring a force onto the blades (Yuksek et al., 2006). Water leaves the turbine from the draft tube, that also serves to be a diffuser, and minimizes the leaving velocity of the water. This is happened to retain an utmost of energy from the running water that aids to make better the efficiency of the turbine. In contrast to the inlet, a 90 turn is made when the water runs from the outlet. The guide vanes (as known as wicket gate) might be tailored to make an effective turbine operation possible for regulating the runner speed by altering the positions of the blades, and for a range of the conditions of the water flow. Power is conveyed out of the blades of the runner towards the turbine shaft in the form of rotation and torque. The shaft of the turbine is openly associated with the generator for achieving electric power.

The shape and size of the blades belonging to a Francis runner are based on their particular speed (refers to box). Higher particular speed refers to a greater construction height of the Francis runner and the lower rotation (refers to Figure 8.11). It means that at the runner outlet, the velocity of discharge must be small, for given power output, to prevent cavitation and simultaneously, the runner must allow a comparatively large amount of water.

Figure 8.11. Two Francis turbines's runners (laboratory size with kW-power range). The right one is designed with a high specific speed, and the left one with a low specific speed.

[Source: https://www.hydroreview.com/world-regions/rehabilitating-the-francis-units-at-chief-joseph/.]

Figure 8.12. Deployment of Pelton wheels at Germany, storage power station Walchensee.

[Source: https://commons.wikimedia.org/wiki/File:Walchenseewerk_Pelton_120.jpg.]

8.2.2. Pelton Turbines

In 1889, a scientist named Lester Pelton has been developed the Pelton turbine (see Figure 8.12) in the USA. This kind of turbine is beneficial for a low volume flow like up to around 40 m^3/s, and for water drop's high heights between 200–2,000 m. A conventional Pelton turbine has resembled the water wheels used in previous times (Abe, 1979; Canadell et al., 2019).

There is a circular disk on the Pelton Turbine which is mounted on the rotor or rotating shaft. A bucket/cup shaped blades are carried by this circular disk known as buckets, situated at an even distance at its circumference (refers to Figure 8.13). Multiple or single nozzles are organized over the wheel. The water jet coming out of a nozzle is moved in a tangential direction to the Pelton turbine wheel's circumference to hit the buckets (refers to Figure 8.4 as well). All nozzles and shape are located over the Pelton wheel, according to the operating needs and the available water head (i.e., the water pressure).

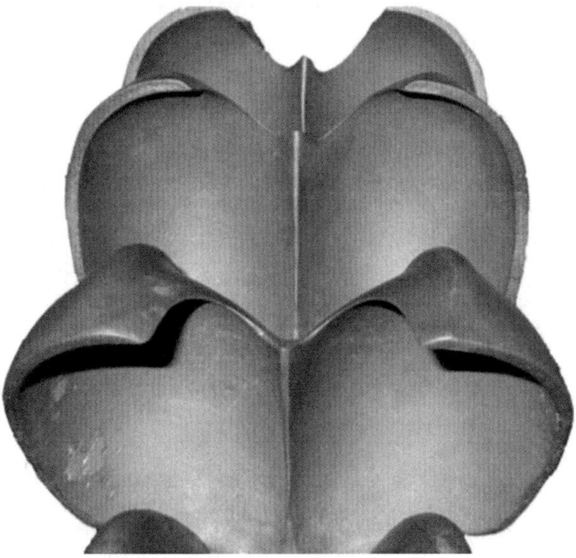

Figure 8.13. Shaped blades along with splitter of a Pelton wheel.

[Source: https://link.springer.com/book/10.1007/978–3-642–20709–9.]

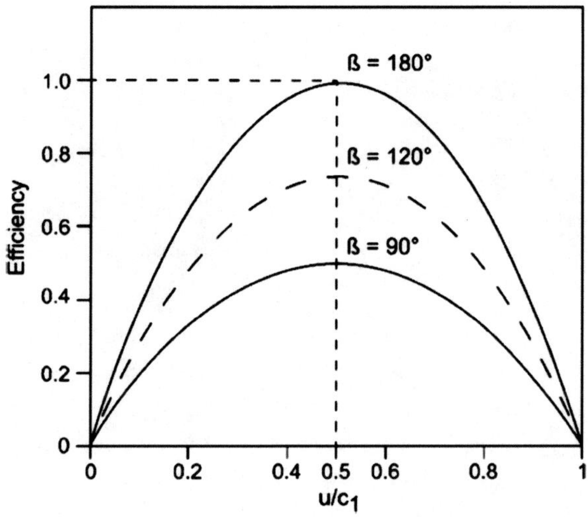

Figure 8.14. The efficiency of a Pelton runner based on the shape and velocities of bucklets.

[Source: https://www.springer.com/gp/book/9783642207082.]

Water jets of the Pelton turbine, also called the Pelton wheel, effect the turbine's buckets (or blades) which causes the wheel to rotate, and produce power and torque. The whole setup that rotates because of the hitting of the water to the blades is known as a runner. Actually, other losses are also occurred despite the profile losses, e.g., aerodynamic drag, mechanical friction, nozzle losses, and limit the thorough efficacy of Pelton turbines to about 90%.

A typical arrangement of a system producing electricity with the help of a Pelton turbine is having a water reservoir located at a significant elevation above the Pelton wheel. Water coming out of the reservoir is flowed by a pressure channel towards the penstock head (Bächtold & Guedj, 2014). After that, through the supply pipeline or the penstock, it is reached to the nozzles from where the water exits like high-speed jets while hitting the Pelton turbine's blades.

Figure 8.15. Kaplan turbine runner.

[Source: http://www.hydrotech-eng.com/equipment/turbines/kaplan-turbines/.]

8.2.3. Kaplan Propeller Turbine

In 1912, a scientist named Viktor Kaplan has been developed the Kaplan turbine in Austria. The rotor seems like a typical ship's propeller having movable rotor blades (refers to Figure 8.15). Their main special characteristics are low heights of the drop until around 25.0 m, even when it could operate by a greater drop height, and flexible rotor blades in consideration of alterations in volume in a range around 1 m^3/s up to 500 m^3/s (Elfarra et al., 2015; Anup et al., 2016).

On the other side, the Propeller turbine is the same as a Kaplan turbine regarding the required volume flow, height of water drop, and also appearance, but differences are also shown. Whereas it is used for a less or more fixed volume of water, in the case of the Kaplan turbine, its rotor blades are not flexible. While, usually, there is more than one kind of turbine in big power plants which transform around all potential energy through turning off and on different Propeller turbines simultaneously.

Despite the water flow path, both Propeller and Kaplan turbines, belongs to the class of reaction turbine as well. Therefore, their working standards are also quite alike to the operation of the Francis turbine. The Kaplan turbines and particularly Propeller turbines usually appear in river power plants since they are the finest of all kinds of turbine to cope with a low height of fall and high water flow.

A higher flow rate from the turbine is required to produce a significant amount of power by small water heights with the help of the Kaplan turbine. The Kaplan turbine is created to adjust the desired high flow rates. The manufacturing of the Kaplan turbine despite the alignment of the blades is very much identical to that of the Francis turbine. As well, the Kaplan turbine possesses a ring of guide vanes over the water entry point of the turbine (Puertas et al., 2013).

In the Kaplan turbine, there could be a way within the rotor and the guide vanes (see Figure 8.7), which is opposite to the Francis turbine that possesses guide vanes openly at the turbine rotor's periphery. The form of the passage is like when the flow which gets in the way in the radial path is directed to flow in an axial path. The central shaft attaches the rotor blades of the turbine. The shaft are joined to the blades using moveable joints; hence, the blades could be able to swivel with respect to the available water's fall height and flow rate.

The shape of the insertion point of water is of a tube of scroll-shaped that covers the wicket gate of the turbine all around. Water is flowed spirals upon a propeller-shaped runner, and tangentially from the wicket gate, making it to spin.

The exit point is a specifically straying draft tube that aids restore kinetic energy and slow down the water. As described before, turbines could be applied in both orientations, i.e., vertically and horizontally. An effective working for a vast range of flow conditions is allowed by the turbine blades and the variable structure of the flexible gate vanes. There is typically around 90% and more efficiencies of the Kaplan turbine, but it may be reduced in too low water height utilizations (Marongiu et al., 2010; Solemslie & Dahlhaug, 2012).

There are no flexible propeller vanes of Propeller turbines. They are applied in circumstances where the water volume's volatility is not abundant. The range of the economically possible Propeller turbines is few kilowatts up to higher than 100 MW.

Pit turbines are the Propeller or Kaplan turbines that are deployed to being bulb turbines. They possess a gearbox. Generally, these turbines possess too tiny capacity or else are in the order of small or micro-power plants because of the availability of a gearbox. The maximum capacity is restricted by the Gearboxes to around 10 MW. Nevertheless, the very common scheme in the MW range and additional is to connect the turbine openly to a generator using a multipole generator with a fixed amount of poles.

Straflo turbines act as axial Propeller turbine in a much compact design. The blades' tips are associated with a ring, where the poles of the generator's armature are attached. The generator armature and turbine runner are a single section. The generator stator is beyond the water channel, associated with the runner's periphery (Laghari et al., 2013; Chen et al., 2018).

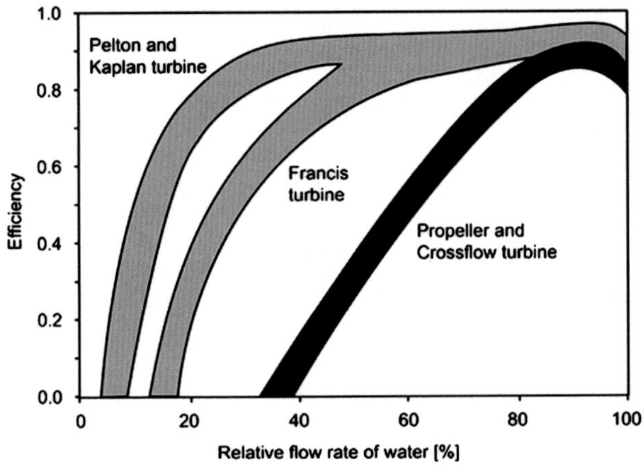

Figure 8.16. Classic efficiency plots for various turbine types (the form might change with respect to the design of the turbine).

[Source: https://link.springer.com/chapter/10.1007/978–3-642–20709–9_5.]

8.3. OPERATIONAL PROSPECTS OF TURBINES

8.3.1. Efficiency

The efficacy of a turbine is based upon the kind of turbine and the water flow (Figure 8.16). The efficacy of Kaplan and Pelton turbines is greater than a broad range of water flow. A distinct optimum is presented by the Crossflow and Propeller turbines. The given-below features must be noted (Svardal & Aggidis, 2016):

- When there is a too minimum flow, then the efficiency of all the turbines is reduced. There should be less water flow in consideration of turbine operation (e.g., for Francis, 30% of rated water flow).

- The highest efficacy of all turbines (rather than the too-small ones) is of the order of 90%, still, this higher efficacy is never over 100% flow.
- The particular speed of a turbine also depends on the form of the efficiency curve.
- For fixed propeller turbines, the efficacy reduces rapidly with the decreased inflow.

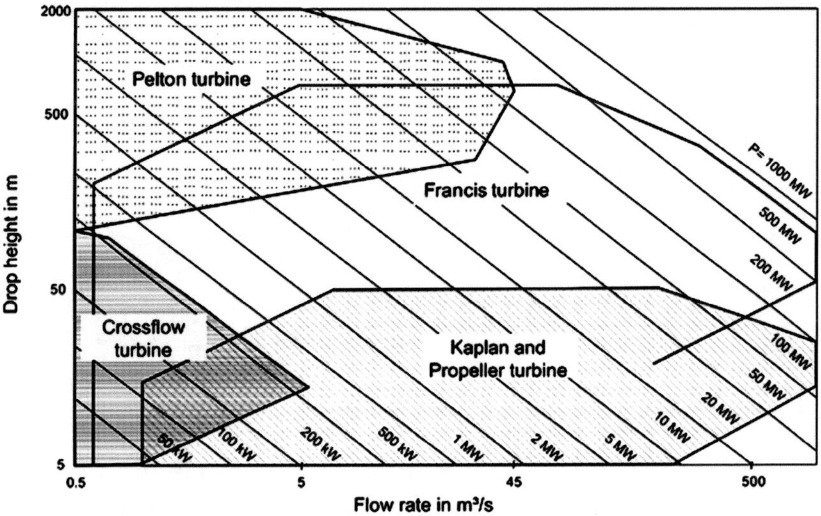

Figure 8.17. Working regions of hydro turbines and their power (over logarithmic scales).

[Source: https://link.springer.com/chapter/10.1007/978–3-642–20709–9_5.]

8.3.2. Choosing a Type of Turbine

If the flow and head of water are recognized, then Figure 8.17 can be utilized to establish a concept that which turbine is appropriate for that specific combination. The indication of the figure is that (Ekama, 2009):

- The Francis turbine is appropriate for medium to high flow and medium to high fall height.
- The Pelton turbine is appropriate for high fall height and low flow.
- The Kaplan turbine is appropriate for medium or low fall height, and medium or low flow.
- There is an overlap between Kaplan and Francis, and Pelton

and Francis turbines. It means that both kinds of turbines are appropriate for such combos of flow and fall height. Nevertheless, the final judgement should be based on all particulars of the real circumstances counting particularly the cost facets.

8.3.3. Two-Block and Three-Block Systems

The central machinery belonging to a pumped storage power plant comprises either 2 or 3 blocks. Although the three-block-system contains turbine, generator, and pump as sole blocks, in the two-block-system, turbine and pump are linked along with a Francis Turbine (Rielly & Gimbun, 2009)

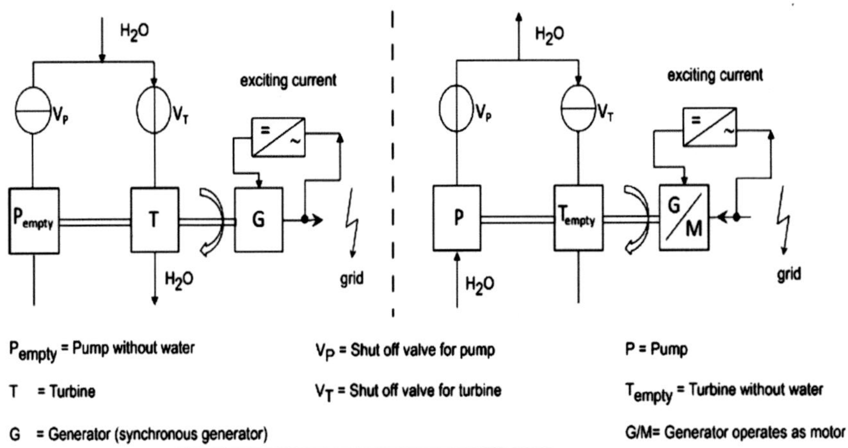

Figure 8.18. Pump and turbine working of a three-block-system.

[Source: https://www.springer.com/gp/book/9783642207082.]

The investment of the two-block system is reduced as there is no need to purchase a separate pump, which is a benefit of the two-block system. While the limitation is the complication of the switching of a turbine within power generating and pumping modes as compared to the working of a three-block system. Other drawback of the two-block-system is the choice of the Francis turbine, operating as a turbine and also as a pump might be a compromise with respect to the efficacy, specifically when the head and flow combination advises some other kind of turbine for getting higher efficiency.

There are three main machine parts of the three-block-system as mentioned (refers to Figure 8.18). Moreover, it possesses two valves, one of these is in front of the pump and one is in front of the turbine. When the turbine is working, the valve that is in front of the turbine is held open and the valve that is in front of the pump is closed. The water flows via the turbine, the turbine switches, and the switching turbine shaft drives the generator (Wong & Sumathy, 1999; Salter et al., 2002). Finally, the generator generates electricity which is provided to the grid. It is easy to turn the machinery from the operation of the turbine to pumping. If the water pumping is demanded by the operator, when the price of electricity becomes low-cost or when the need for electricity reduces, then the valve is closed which is ahead of the turbine and after that, the turbine is vacated through blowing compressed air. When a little water has remained to the turbine, it serves being a brake and the remaining water is also evaporated while function as a pump which harms the blades of the turbine. Within the pump, when the compressed air has completed its operation, it is moved away and the water is inserted into the pump from the downward side by unclosing the valve in front of it. At the outlet of the pump, with the increase of the water pressure farther the pressure of the water needed for water pumping is reversed to the height of the storage, then the valve ahead to the pump is unsealed. After that, the lower-level water is moved up to the upper level. The pump and the common shaft within the turbine move in the same direction. It means that even the working of the plant gets reversed, the direction is not turned around (Carta et al., 2004; Shallcross, 2013).

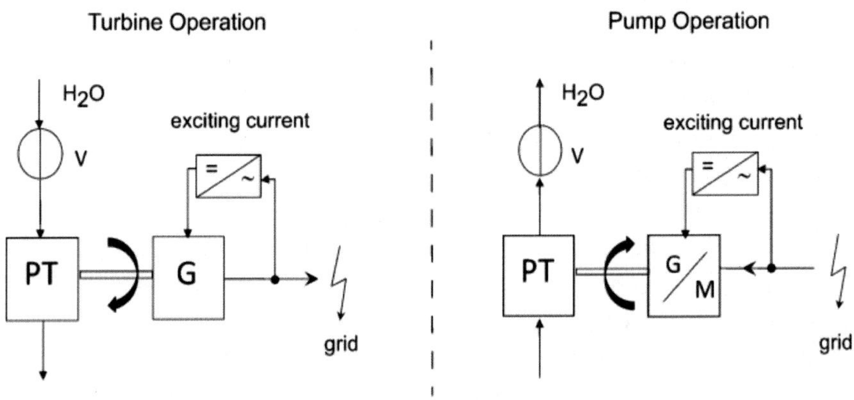

Figure 8.19. Pump and turbine working of a two-block-system.

[Source: https://www.springer.com/gp/book/9783642207082.]

In the pumping operation, as well as known as motor operation, hither, the generator operates as a motor if electricity by the grid is consumed and switches the shaft for the purpose of pumping. The turbine, that is associated with the shaft as well, is operated along empty. When the mode is turned again, from the mode of pumping to the turbine working mode, also known as the generator operation, the valves shutting is completed the other way. Initially, the valve onwards the pump must be closed. In this way, the water within the pumped is taken out and directed back to the lower reservoir through the compressed air. When the pump is vacant, another valve onwards the turbine is active to be turned on (Linnemann et al., 2003). When the valve is opened, first the water forces the air from the turbine out, and after that, the water runs through the turbine, then the electricity is regularly produced by the generator. While the turbine changes the shaft in those modes, the pump also operates along vanact. Even though there are two valves of the three-block system, it cannot happen that both the pump and the turbine do their task simultaneously.

However, there is a different operating scheme of the two-block system. As given above, one single unit operates being a turbine and pump both (refers to Figure 8.19). When the upper level is reached by the water, the Francis runner acts as a turbine and it operates being a pump, when water flow moves in the other direction. Therefore, just a single valve is required in this kind of configuration. The turning from pumping operation to the turbine operation operates a little different in contrast to the three-block-system. When the operation is switched, first the valve is closed and the pumping is halted. The valve must be closed till the pump has stopped switching completely, as it must switch into the other direction throughout the turbine operation. After that, the valve is again turned on and the pump is started its operation being a turbine, switched from the upper-level of water producing electricity. The valve must be closed again for turning back to the pumping mode till the runner stands to halt its rotation. After that, the runner is moved in the reverse direction directed from the generator operating as a motor. The valve is turned on when at the valve, the pressure of water within the pump is greater than the static pressure from upper water. No compressed air is needed for a two-block system for clearing the pump or turbine (Arnulfi & Marini, 2006; Majewski & Yager, 2016).

In a two-block system, each turning indicates that the entire block system must be driven up and down further. Previously, this used to require more time as compared to for a three-block system. Their operation mode can be changed by the modern power stations having two-block-systems in around 2–3 min (Braitsch & Haas, 1996; Momčilović & Odanović, 2019).

8.4. USE OF OCEAN ENERGIES

The technology in consideration of harnessing the water power on land has already discussed in the previous chapters of this book. Moreover, the opportunity is available for generation of power from the water of oceans. The technology participating in the conversion and utilization of the power of the ocean will also be discussed in this chapter. For this purpose, different technologies can be categorized into five major classes (Aymerich et al., 2015):

- Tidal power plants;
- Ocean current power plants;
- Wave power plants;
- Ocean thermal power plants;
- Osmotic power plants.

Rather than tidal power plants, all mentioned above technologies are under the development and research phase of prototypes.

8.4.1. Tidal Power Plants

The gravitation of the sun and the moon generates one of the most compelling scheme of natural energy globally. High and low tides are caused by this movement according to the earth, which occurs two times within a period of 25 h. The potential energy of water is changed by the movement of the falling and rising sea level that can be transformed into electricity from the power plant operation. A dam wall is formed, for using the potential energy, which encloses a specific volume of seawater within an artificial bay facilitating the need for a reservoir for a hydropower plant. With the rise of the tide, the water gets in the reservoir via turbine which generates electric energy until there is high seawater within the reservoir, high up to outside water level (Horsburgh & Wilson, 2007). The reverse process starts, when the tide is low outside the reservoir, and the water within the reservoir get off into the sea via a turbine. Practically, for around two hours, there is a pause within both processes, as a least of water levels should be available to encompass the energy losses within the turbine. There is not a constant capacity of power generation of the running water (refers to Figure 8.20) because the water flow pressure differs with the saturation state of the reservoir.

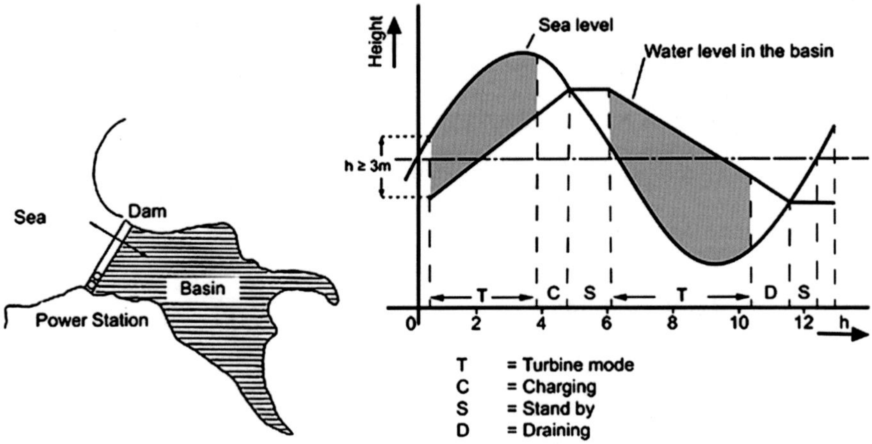

Figure 8.20. Operation demonstration of a tidal power plant.

[Source: https://link.springer.com/chapter/10.1007/978–3-642–20709–9_6.]

Consequent to these variations, it cannot be utilized permanently to encompass a constant load due to the interruption which is a limitation of utilizing this kind of energy. Whereas, the advantage of these systems is the power generation predictability as the amount of water flow and time can be predicted correctly (Woodworth & Blackman, 2004).

The quantity of electric energy that can be transformed from the tides relies upon the sea level amplitude. For economic and efficient usage, tidal energy must exhibit a height of a minimum of three meters starting from its average level. It indicates that particularly the zones having extremely large tidal ranges are appropriate for economic and efficient usage of tidal power.

As a result, the often used turbines are the axial flow turbines, e.g., Kaplan turbine that utilized for tidal power stations.

8.4.2. Tides Formation

Tides are formed when the moon and the earth are attracted to each other because of their own gravitational pulls. An attraction is wielded by the moon on all things on the earth to take it nearer and the earth does the same. The capability of the earth is to keep up everything from the attraction of the moon, rather than for water. As the water upon earth is in motion always, the earth cannot dominance it strongly, and consequently, the moon is capable to attract it to some amount. Daily, there are two low tides and two high tides (Woodworth et al., 2002). The motion of the level of the ocean is constant which is from the high tide towards low tide, and again then back to high tide. The time is approximate around 12 hours and 25 minutes within two successive high tides.

As the location of the moon and the earth according to the sun variates during the whole year, a variable quantity of energy is also produced by the tidal power plant as spring tide, while the neap tide changes in the routinely sea level and provide different quantities of potential energy. The gravitational pull of the sun and moon are mixed for the new or full moon. With this combo, the low tides are too low and the high tides are too high. These situations are called spring tides. Nevertheless, there is no connection of this to the spring season as it might conclude by the name. When the sun, the moon, and the earth are in a line, they appear. There is a contribution of the gravitational forces of both the sun and the moon to the formation of the tides. During the new and the full moon, spring tides appear (Pingree, 1978; Flexas et al., 2015).

The other tidal situation occurs while neap tides. The pulls of the moon and sun operate at right angles for the moon's quarter stages, hence, cancel the effect of each other. It's consequence is in a lowest net pull performing on the water and as a result, there is a less difference within low and high tides that are too weak tides.

8.4.3. Existing Tidal Power Plants

Presently, just two commercially rated tidal power plants are working globally. One of them is in Annapolis Royal, Canada, and the other one in La Rance, France.

In 1984, a significant tidal power plant has been created in Canada over the southeast coast town of Annapolis Royal. Single directional turbines are possessed by this plant that has the ability to produce 20 MW power. The Annapolis River is obstructed with a rock-filled dam for making a reservoir. However, the estimated potential for this location is to produce up to 5,000 MW of power, but because of the colossal influence that the reservoir increase and the construction would have upon the regional and local environment, the plan has been restricted (Gao et al., 2010).

In 1966, the initial and still the world's biggest tidal power plant is constructed which is situated in La Rance. This location serves a variation in tidal height, up to 8 m in the mouth area utterly regularly. There are twenty-four hydro turbines in this plant having a general capacity of 240.0 MW that generate electricity of 600.0 GWh per year.

Other tidal power plants operate with a much little capacity. In the world, there are about a hundred areas in which tidal power plants can be made to operate efficiently.

The biggest tidal power plant is being built up in the south of Seoul, Sihwa in South Korea. Ten turbines having 26 MW capacity of each will be contained by the power plant, consequenting to the accumulative capacity of the plant reaching 260 MW. Indeed, the idea of the larger bigger tidal power plant is planned in Great Britain, at the edge of Britain's longest river, Severn, that runs into the Atlantic Ocean within Wales and England. The creation of a dam between Bristol and Cardiff would have covered in the proposal, more specifically a formation of a 16 km long dam. However, the project is terminated by Great Britain in 2010 which claims that it would be too expensive (Blomster et al., 2002).

The truth is that the high initial investment required for the power plant is the main disadvantage as compared to other power generation choices.

The environment is also concerned by other negative view of tidal power plants both in an aesthetic and a natural sense. There is a high effect of making of a concrete dam and a big reservoir on natural life in the water and on land. However, the calculable and safe supply of electricity independent of fossil fuels and without any emission is the principal motivation for making tidal power plants.

8.4.4. Ocean Current Power Plants

The terminology ocean current is the ocean water's mass flow. For power generation, when this mass movement is used from its kinetic energy, it gives for ocean current power plants. Such current power plants may be messed up with tidal power plants because of a similar operating standards, but there is a slight difference within them (Rauf et al., 2015). The mass movement of water might be either an uninterrupted movement which is not limited to a narrow region but is global, similar to the gulf, or it can be a small regional movement whose nature is periodic (similar to tides). There are quite diverse reasons for such movements. The friction of wind is one of the notable reasons that it flowing over the free water surface and compels the peak by water's sun layers to be in motion. The upwelling phenomenon is another cause of the movement of water, where the nutrient-poor and upper warmer layers are interchanged with the nutrient-rich and lower cooler layers which makes a specific vertical current within certain layers, whereas the horizontal streams are primarily caused by the movement of the wind (Cavaleri et al., 2015). The dissimilar warming up of the ocean water through the sun is another sense of the movement of ocean water. This result increases a vertical temperature gradient consequently is largely vertical streams. Moreover, the motion is caused by the salinity and density of water as they fuse with the encompassing water that does not possess these qualities and differ locally. The topography of the ocean governs largely the available water streams, i.e., deeper and higher bottoms generate vertical and horizontal water currents. Other forces that influence the current are considered to be the centrifugal force produced by the earth rotation, collision with landmasses, temperature, and friction.

There is a broad variety of technological approaches for transforming the water current for the electricity production. The initially commercial-oriented scale ocean current power plant is the Italian ocean current power plant known as Kobold (Figure 8.21). The manufacturing of this plant has been completed in 2001 and is situated in the Mediterranean Sea close to, Italy. There is a floating part on this tripod machine that incorporates the generator and alternate equipment. There are many underwater parts of this machine that made up of three propellers, a shaft, and a frame for such underwater construction (Kok et al., 2010).

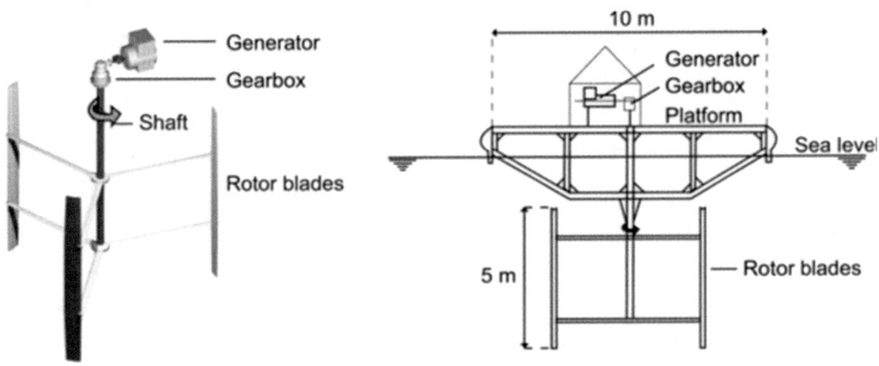

Figure 8.21. Diagram of Kobold experimental ocean current power plant.

[https://www.springer.com/gp/book/9783642207082.]

The three propellers are somewhat alike a vertical axis wind turbine rotor. As the water density is much greater than the air density, the rotor dimensions of ocean current power plants may be lower by the same power comparable to wind energy rotors. The streams of water flowing can be used by the power plant in either direction because of the shape of the rotor. The diameter of the rotor of the Kobold plant is six meters, but it demands a minimum 2 m/s flow velocity to initiate turns. The highest power generation capacity of it is 130 kW.

Another creative experimental current power plant is a stingray that is situated on the sea bottom close to Yell Sound away from the Shetland Islands in the sea nearby Scotland. There is a four-legged frame of this construction having a holder for both wings which propel with the current that utilizes vertical streams. The starting speed of the motion of the wings is 2 m/s which is the same as that of the Kobold power plant, while its size is 150 kW as well (Tai et al., 2013).

In 2003, alterations of the prior notions for ocean current power plants are exhibited, when the British Seaflow has been constructed as a test rig close to the Lynmouth village in the Bristol Channel, by the British company Marine Current Turbines Ltd. This formation is alike a converter of the horizontal axis wind energy because it possesses a rotor along with two blades. The complete power plant is constructed underwater which transforms horizontal water current to electricity. The generator and rotor are hooked up to a lifting collar, and a top part is raised to sea level. Until 2009, the test rig is worked, until a further commercial technology, Seagen has been introduced (Figure 8.22). That is the same model, which is designed by Marine Current Turbines Ltd as well. This power plant possesses a holder along with two rotors. It is deployed on the coast of Strangford Lough, Northern Ireland, having a 1.2 MW capacity (Kapoor et al., 2021; LaCroix et al., 2021).

There are two benefits of this kind of plant. First one is that the ocean energy can be utilized more accurately by placing the rotor in the layer using the very powerful current. Moreover, the collar can be raised above sea level that makes easier the service-related task and repair. The drawback is the large investment cost because of the construction. A least flow velocity of 2 m/s, is required by the sea flow, for initiating to produce power and it possesses 350 kW of rated capacity.

For Seagen, the least speed of water must be 2–3 m/sec. There is a vast area of spots in the oceans that present current energy, thus to construct such power plants, thousands of kilometres of coasts can be utilized. But various types of challenges must be kept in mind while planning the manufacturing of ocean current power plants, farther and deeper from the coast (Marcus et al., 2016). These plants, particularly those at the prototype phase, usually require maintenance and repair and, hence one significant consideration should be the easy access to the power plant. Furthermore, for power transportation to the shore, sub-marine cables must be well-structured. Moreover, one of the prime concerns is the increase of need of starting investment for manufacturing in depth. Considering above-mentioned aspects, usually, it is recommended to construct new power plants in coastal areas instead of selecting a site deep into the open sea.

Figure 8.22. Seagen ocean current power plant along with graphic inset to display operation.

[Source: https://www.power-technology.com/projects/strangford-lough/.]

8.4.5. Wave Power Plants

The continuous movement of the wave on the water surface is used by the wave power plants that are occurred by wind and tidal energy utilized by wave power plants. Further precisely, wave energy is largely an indirect kind of wind energy that induces motion on the surface of water stretches. Following are the four different standards to transform wave power into electric power:

- Wave energy turned into potential energy;
- Wave energy transformed into air pressure;
- Waves trigger the hydraulic flap's oscillating movement;

- Various altitudes of water surface converted into mechanic energy.

The initial prototypes of power stations that transform wave energy into pressed air are constructed in the 1980s in the Trivandrum in South India, and in the 1970s in Norway. Both prototypes have been working successfully for some years, but are now obsolete. The wave water is flowing into a case, that is opened on the seaside and closed by the coast. The volume is split into two subvolumes by a lip within the case. Behind the lip, when the water is in the subvolume, the inside air is pressured. An air propeller is driven by the pressured air that causes to turn the generator. The refelection of the wave is on the rear wall of the case and the air is inhaled into the case from the outside when the water exceeds the sub volume from the rear of the lip. Controlling valves makes sure that the air must always be moving in the alike direction through the wind propeller (Fabian et al., 1975).

Figure 8.23. LIMPET wave power plant (image captured by courtesy of Voith Hydro Wavegen).

[Source: https://news.climate.columbia.edu/2010/05/19/limpet-land-installed-marine-powered-energy-transformer/.]

The commercially made and working power plant situated in Scotland, named LIMPET (i.e., Land Installed Marine Powered Energy Transformer), is dependent on the precept of transforming wave energy within compressed air (refers to Figure 8.23). The overall capacity of this power plant is 500 kW. The position of the power plant in a crevice is such that a slanting segment of the roof goes into the water, whereas the other segment of the roof is aligned above the land. There are two wind turbines in the land-facing wall which converts wind into electricity (refers to Figure 8.24). The precept of the communicating vessels is the basis of the consumption of energy, which signifies that the levels of water outside and inside of the building will seek to coordinate with each other. Because of the wave movement when the outside level is greater than the inside level, water will get in the reservoir and compress the inside air of the building (McMillan et al., 1987).

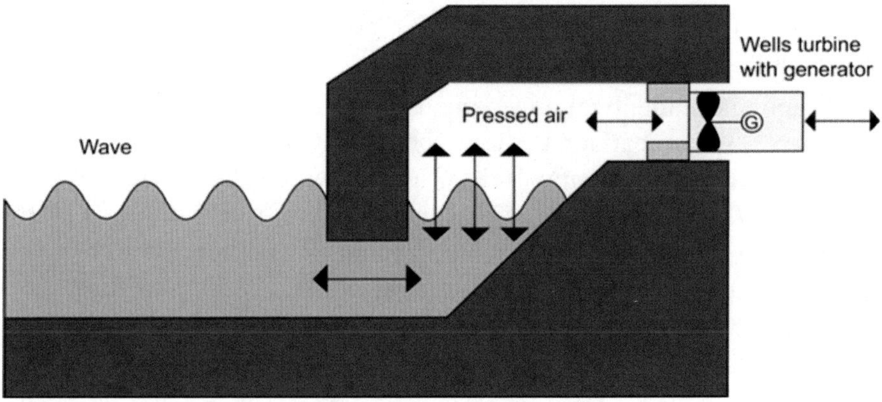

Figure 8.24. Illustration of the LIMPET power plant.

[Source: https://link.springer.com/book/10.1007/978–3-642–20709–9.]

The turbine is driven by this compressed air and generates electricity. The water level inside the slanting roof will also sink when the water level outside sinks again, and air enters inside via the turbine. The well-known Wells turbines are deployed in the LIMPET power plant (i.e., 500 kW rated power). With the rotation in the same direction, the air is allowed to invade in both directions from the turbine.

Hither, water power is used indirectly instead of directly. In actuality, hither, wind energy is utilized that appears from the various water levels induced by wave movements.

The second rule of the wave power plants's working is by transforming wave energy to potential energy. In actuality, wave energy and kinetic energy are same that requires to be transformed into potential energy through lifting the water onto an upper level. The first wave power plant that has done this, named Tapchan, stands for the tapered channel. In 1984, it is built in Norway. Waves reach an increased and tapered channel in this plant, where there is a narrow cross-section regarding water flow channel in consideration of inflow (Ray et al., 1987; Lai, 1997). After that, the water is gathered in a reservoir from where it retires in a controlled way via a simple propeller turbine into a pipe again into the ocean. The opening of the channel of the Tapchan plant is 60 meters broad from where the water flows in. The surface of the reservoir's last segment is around 3m upper than the sea level, and the reservoir is around 7 meters deep. The capacity of the plant is that it produces 350.0 kW electric power through wave energy. The yearly generated energy is approximately 2.0 GWh. Already the Tapchan power station is not operating.

Wave Dragon is an identical power plant that turns wave energy towards potential energy. In comparison to the Tapchan, that is a fixed plant at the coast, there is a floating construction of the Wave Dragon. It has been deployed in Denmark in 2003 on the coast of Nissum Bredning (refers to Figure 8.26).

The Scottish experimental machine Pelamis is another way to turn wave energy into electricity, located in Portugal. Pelamis refers to the class of water-living snake which presents the idea for the name of the building of snake-shaped of this plant.

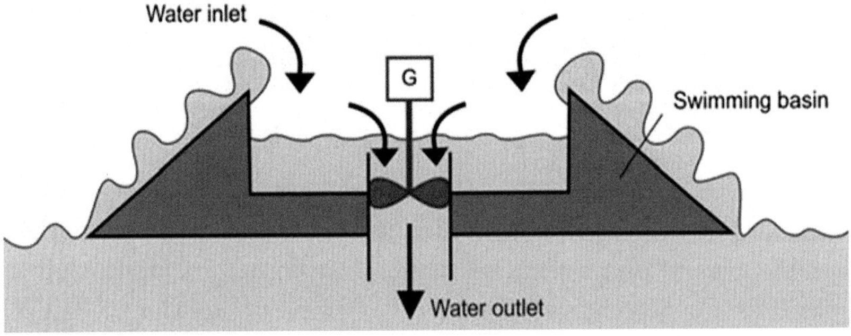

Figure 8.25. Demonstration of the Wave Dragon experimental power plant.

[Source: https://link.springer.com/chapter/10.1007/978–3-642–20709–9_6.]

This plant has been begun to operate in 2006. It has three, 140 meters long, tubes of metal that create the main part of this plant. The third rule of consuming wave energy is using in this plant: the turning of various altitudes of the water surface into mechanical energy and eventually into electricity. Many hollow cylinders are involved in each of the tubes that are associated with the hydraulic pump through joints. On the propelling of the waves by the snake-shaped power plant in various spatial directions, the oil is pumped by the hydraulic cylinders from the oil pump of high-pressure towards pressure tanks from where the oil is not able to move back. Hither, the pressure is hoarded and the compressed oil leaves the pressure tank using an expander (Ray et al., 1987; Lai, 1997). The pressure is converted into mechanical rotation energy by the expander which gives power to an electric generator and minimizing the oil's pressure and turning it back towards the high-pressure oil pump, next the process iterates. For the case of all offshore deployments, a submarine cable carries the energy to the land, from the ocean, and then into the electricity system. A maximum of 2.26 MW power can be delivered by the three tubes, i.e., each 750 kW.

In 2006, another original of its type wave power plant is constructed in Denmark, named Wave Star, on the northwest coast. This plant is constructed having a body with floats connected to either side. The shape of these floats are similar to half balloons and their diameter is 1m (refers to Figure 8.26). They hover onto the water surface and transfer the motion to the holder, from where the movement is transformed towards electric power. The Wave Star begins to generate energy from a 5 centimetres wave height difference, and the capacity of its power production is until 5.50 kW. However, the rule on which the power plant is based and that of Pelamis is the same, as the variating water levels also run oil pumps.

Figure 8.26. Wave Star (image captured by courtesy of Wave Star Energy).

[Source: https://renews.biz/42407/wavestar-future-on-knife-edge/.]

The fourth rule of using wave energy is performed by a certain kind of wave power plant where the waves get in a chamber. A flap in a chamber exists. The flap is activated by the water when getting in the chamber, by push down its front side, which reflects on the chamber rear and runs back. The backside of the flap is pushed down by this movement and the water runs back to the ocean. The motion of the flap generated by the water is transported to hydraulic cylinders. Oil is pumped by these cylinders into pressure tanks. The working is alike the working of the Pelamis plant that is explained before. This kind of one of the prototypes is available in Yagishiri, Japan (Shrestha & Choi, 2020).

There are three units in this power plant having a total of 450 kW electric power. Each unit is a case having about 8 m width, 13 m length, and 10 m height. There are different challenges that the usage of wave energy is facing. Wave power plants are prone to tough weather conditions; thus, few experimental plants are already ruined from the strong storms and tides forces. Repair and maintenance are expensive, and still, there are slight experience data. Although, it is difficult to finance wave power plants because banks ponder to grant credits for such insecure projects.

8.4.6. Ocean Thermal Power Plants

There is a vertical thermal gradient of water bodies. The water of the subsurface and ocean surface is comparatively warmer than the fluid of deeper layers. The temperature difference of various depths can be transformed openly into electricity as well by an OTEC (Ocean Thermal Energy Conversion). According to the thumb rule, the water temperature variates by 5.0 K subsequent to every 100 m depth. Figure 8.27 illustrates a conventional OTEC power plant where there is a closed circulation system involving a fluid operating medium like organic fluid or ammonia, which is transmitted from system's deeper to shallow layers and returns back to the deeper layers again. The selection of operating medium has relied upon the water temperature in the deeper and shallow layers of water (Marcus et al., 2016).

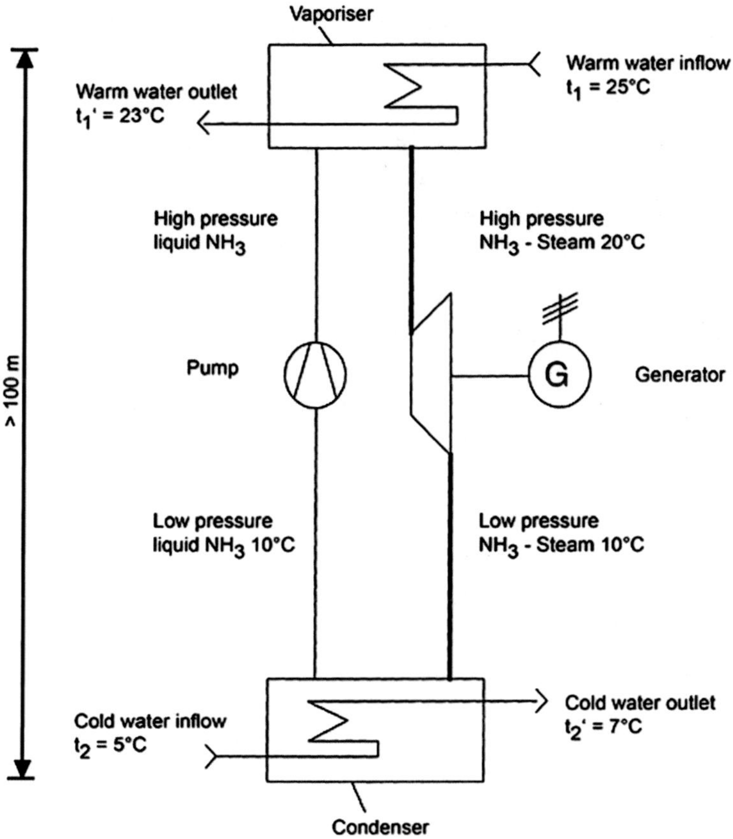

Figure 8.27. Diagrammatic image of an oceanic thermal power plant OTEC.

[Source: https://www.springer.com/gp/book/9783642207082.]

The circulation of operating fluid is not made by itself, hence, there is a need for a pressure pump to pump the channel from deeper levels to the shallow levels that produce a greater pressure in the shallow portion of the system as well at the same time. Here, in the shallow portion, the medium arrives at a vaporizer where it turns evaporated. For instance, the 25.0° C temperate surface water gets in the vaporizer and transfers hotness from ocean water to the operating fluid for their evaporation. Hence, during leaving the vaporizer, the temperature of the water outlet is reduced to few degrees. The delivered energy supports the operating medium to get evaporated. The vapour infiltrates the steam turbine which triggers a generator. The temperature and the pressure of the vapour in the turbine are minimized. While exiting the turbine, vapour will move to the condenser. It is situated at the lowest area of the power plant, which might be 100 meters and more. For example, the temperature of incoming ocean water is 5.0° C, which causes the vapour condenses, and when leaving, is a minor degrees warmer. Now again, the vapour is transformed into a liquid, and for restarting the cycle, it is forced in with high pressure into the system (Shrestha & Choi, 2020).

Briefly, in the working of an OTEC power plant, there is a reduction in pressure, and the cool liquid medium flows in the upper direction from the condenser. The pressure is increased by the pump. After that, the medium approaches the vaporizer where when getting heat from warm water, it is evaporated. The evaporated, the warm operating medium having higher pressure, triggers the turbine where temperature and pressure are minimized. The generator is driven by the turbine. Ultimately, the medium approaches the condenser in which again it transforms into a liquid state. Generally, the process is similar in every steam power plants, just the operating medium is changed from steam and water.

Same as other offshore power plants, a submarine cable is also required by this plant for the transportation of produced power to the land, from the ocean. The order of three percent is the efficacy factor of an ocean thermal power plant, which is one of the dominant obstacles for its use at a commercial scale.

The benefits of ocean thermal power plants are absolutely emission-free power generation. There is no consumotion by the power plants of all the area of the oceans' surface as well and do not require much space. The existence of energy is comparatively secure and nonfluctuating as well. The most predominant benefit of such systems is that the ocean water is not cooled down right after the sunset, which consequently is able to produce power generation after sunset in the last evening hours, exactly at the time when the energy is demanded by several households and additional power is demanded by consumers (Pingree, 1978).

Whereas, still, the ocean thermal power plants are not constructed at an economical level. Some practical issues are incorporated into their work. For example, limpets and other molluscs that are connected to the water openings of the power plant, limit the temperature-dependent circulation and minimize the power generation efficiency. Moreover, another drawback is that as these plants are not really close to the land and extend deep into the water, thus, they are not very well reachable for repair and maintenance. Albeit they are held in the sea bottom, thus, there is a possibility that mighty storms may damage the machines. Complications are involved in these aspects with the working of warmth projects of the ocean. Some of the investors are ready to offer enough funds for expensive construction and experiments. The efficiency is not really high, because 80% - 90% of the generated energy is required for the purpose of pumping. In 1981, an experimental plant is built in Japan. The power generation capacity of this plant is 100 kW, where 90% is used by the pressure pump utilized to flow the system. In the years from 1993 to 1998, another ocean thermal power plant is operated successfully in Keahole Point in Hawaii. Especially for the warm water surface, particularly in late summers, the plant produced 250.0 kW, from which 200.0 kW are required for the pressure pump. Still, the experience is restricted, hence, more development and research are required to make better this type of power plant (Rojstaczer & Agnew, 1989).

8.4.7. Osmotic Power Plants

The osmotic power plant utilizes the energy of osmotic power, as the name indicates, it is the energy that exists because of the salinity gradients. The procedure of directed motion of the molecules from a selective membrane is called osmosis. Molecules of a particular kind are allowed to penetrate the membrane in both directions in a perfect selective membrane, for example, water molecules, whereas other elements, e.g., ions (from salt molecules), are not able to pass through the membrane. A certain pressure gradient is formed because of this directed selective diffusion, which is known as osmotic pressure (Ray et al., 1987; Lai, 1997)..

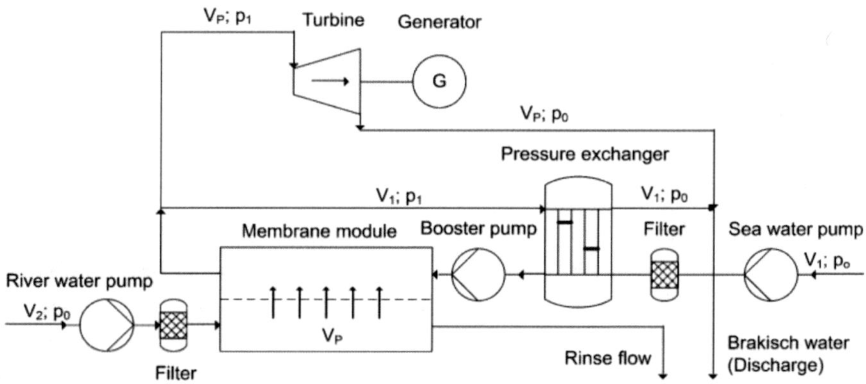

Figure 8.28. Image of an osmotic power plant.

[Source: https://link.springer.com/book/10.1007/978–3-642–20709–9.]

There is a natural tendency of the mixture to balance the concentrations on the contact of the fresh and saltwater. When the fresh water is mixed with salt water on a selective membrane, the outcome is a net volume circulation of water molecules out of the freshwater side towards the saltwater side of the membrane, through dilution of the saltwater (that is the principle of osmosis). The process terminates when hydrostatic pressure and osmotic pressure, that is the pressure of water induced by gravity, are high equally on two membrane sides. Thermodynamic equilibrium is reached at this point.

The diagrammatic image of an osmotic power plant is shown in Figure 8.28. The difference in salinity within fresh and seawater can be utilized for producing power based on the precept of osmosis.

River water and seawater must be separated to prevent the mixture of these waters while inlet to the power plant. Hence, a gap between the river water and the seawater intakes is needed. For osmotic power plants, appropriate regions can be determined at the cavity of rivers where the saltwater coming out of the sea and fresh water coming out of the rivers are about to get mixed (Lee & Ostriker, 1986).

The main elements of osmotic power plants are pumps, inlet and outlet for river water and seawater, pre-treatment (for instance, filters), pipes, membrane modules, and a system of pressure exchanger for strongly efficient recovery of energy (McMillan et al., 1987).

Pressureretarded-osmosis (PRO) is the technical procedure of energy transformation based on osmosis. In PRO, there is a pre-pressurization of the seawater V_{-1} to somewhat half the osmotic pressure, since the river water V_{-2} and the sea V_{-1} water are carried along with each other to the membrane modules. The pressure of permeate volume flow V_{-P} is raised osmotically across the membranes that are utilized to drive the turbine in osmotic power plants which are illustrated in the figure.

The mixing of river and seawater is a component of the natural global water cycle. The surface seawater is evaporated by the sun; hence, the top layer salinity is increased from where the evaporation occurs. The evaporated water turns back as rain and runs into the rivers, whereas the salt continues to be in the seawater. Therefore, the salinity gradient within the river and seawater at any area is reliant upon the solar radiation intensity. The chance for non-fluctuating power generation is one of the dominant benefits of osmotic power plants, as compared to traditional hydropower plants (Rojstaczer & Agnew, 1989).

Figure 8.29. Prototype of osmotic power plant in Hurum/Norway, i.e., started November 2009 (image captured by courtesy of Flickr (Statkraft)).

[Source: https://www.power-technology.com/projects/statkraft-osmotic/.]

The mean salinity of the global ocean water is 3.50% which causes up to 27.0 bar osmotic pressure. This pressure is equivalent to a 270 m water head; thus, a potential of power generation exists. Nevertheless, in PRO, nearly just half of the pressure is utilized for energetic gains, as the power plant is operated typically at half of the osmotic pressure (i.e., the pressure of the entering seawater). This is because of the properties of the membrane power, which arrives at the utmost at these working conditions. For an osmotic power plant, under these conditions, an installed capacity of about 0.60–0.80 MW is facilitated by a 1 m^3/s volume flow of river water. Generally, the volume flow ratio within sea and the range of the river water is 1–2. (McMillan et al., 1987)

For osmotic power plants, theories and plans are completely obsolete, still, the commercial use of this idea is limited because of the non-existence of efficient membranes at old-ages. The earliest osmotic power plant has been built in Norway (refers to Figure 8.29) having around 4 kW installed power. The membrane area of osmotic power plants is around 2.000 $meter^2$. The goal of the prototype is to assess the especially constructed PRO-membranes with conditions of regular operation. The EU, the Norwegian company Statkraft, and Norway mainly supports the project.

Table 8.1. Status of Tidal Power Plants

Name	La Rance	Kislogubsk	Jiangxia	Annapolis Royal
Location	France	Russia	China	Canada
Power	240 MW	0.4 MW	3.2 MW	18 MW
Construction year	1966	1968	1980	1984

The power generation's technical potential globally through osmotic power plants is predicted to be about 5,200.0 TWh per year, but in Europe, it to be around 400.0 TWh per year. Moreover, the ecological potential drives to the technical rules of the energy transformation, the ecological limitations of water eradication into account. As compared to the technical potential that value is quite less.

Table 8.2. Ocean Current Plants' Status

Name	KOBOLD	Stingray	Seaflow	Seagen
Location	Italy	Great Britain	Great Britain	Great Britain
Power	40 kW	150 kW	350 kW	1.2 MW
Year of construction	2001	2002	Out of operation	2010

The chief task for future development is to proceed with the advancements of membranes (salt rejection, membrane power etc.) because the presently existed membranes are not adequately good still for the working of a commercial power plant (Rojstaczer & Agnew, 1989).

REFERENCES

1. Abe, K. (1979). Size of great earthquakes of 1837–1974 inferred from tsunami data. *Journal of Geophysical Research: Solid Earth*, *84*(B4), 1561–1568.
2. Anderson, S. D., Nogueira, M., & Marques, F. L. T. (1993). Tide-generated energy at the amazon estuary: the use of traditional technology to support modern development. *Renewable Energy*, *3*(2–3), 271–278.
3. Anilkumar, T. T., Simon, S. P., & Padhy, N. P. (2017). Residential electricity cost minimization model through open well-pico turbine pumped storage system. *Applied Energy*, *195*, 23–35.
4. Antonio, F. D. O. (2010). Wave energy utilization: A review of the technologies. *Renewable and Sustainable Energy Reviews*, *14*(3), 899–918.
5. Anup, K. C., Lee, Y. H., & Thapa, B. (2016). CFD study on prediction of vortex shedding in draft tube of Francis turbine and vortex control techniques. *Renewable Energy*, *86*, 1406–1421.
6. Armaroli, N., & Balzani, V. (2016). Solar electricity and solar fuels: status and perspectives in the context of the energy transition. *Chemistry–A European Journal*, *22*(1), 32–57.
7. Arnulfi, G. L., & Marini, M. (2003, January). Analysis of transient performance of a compressed air energy storage plant. In *Turbo Expo: Power for Land, Sea, and Air* (Vol. 36878, pp. 63–70).
8. Arnulfi, G. L., & Marini, M. (2006, January). Management Strategies for a Compressed Air Energy Storage Plant. In *Turbo Expo: Power for Land, Sea, and Air* (Vol. 42398, pp. 133–142).
9. Aymerich, I., Rieger, L., Sobhani, R., Rosso, D., & Corominas, L. (2015). The difference between energy consumption and energy cost: modelling energy tariff structures for water resource recovery facilities. *Water Research*, *81*, 113–123.
10. Bächtold, M., & Guedj, M. (2014). Teaching energy informed by the history and epistemology of the concept with implications for teacher education. In *INTERNATIONAL handbook of Research in History, Philosophy and Science Teaching* 115 (2), 211–243.
11. Bahrami, S., Tribes, C., Devals, C., Vu, T. C., & Guibault, F. (2016). Multi-fidelity shape optimization of hydraulic turbine runner blades using a multi-objective mesh adaptive direct search algorithm. *Applied Mathematical Modelling*, *40*(2), 1650–1668.

12. Barnes, J. E., & Hernquist, L. (1992). Formation of dwarf galaxies in tidal tails. *Nature*, *360*(6406), 715–717.
13. Bergant, A., Ross Simpson, A., & Vìtkovsk, J. (2001). Developments in unsteady pipe flow friction modelling. *Journal of Hydraulic Research*, *39*(3), 249–257.
14. Bergant, A., Simpson, A. R., & Tijsseling, A. S. (2006). Water hammer with column separation: A historical review. *Journal of Fluids and Structures*, *22*(2), 135–171.
15. Blomster, J., Bäck, S., Fewer, D. P., Kiirikki, M., Lehvo, A., Maggs, C. A., & Stanhope, M. J. (2002). Novel morphology in Enteromorpha (Ulvophyceae) forming green tides. *American Journal of Botany*, *89*(11), 1756–1763.
16. Braitsch, W., & Haas, H. (1996). Turbines for hydroelectric power. *Landolt-Börnstein, group VIII: advanced materials and technologies*, *3*, 493–604.
17. Braitsch, W., & Haas, H. (1996). Turbines for hydroelectric power. *Landolt-Börnstein, group VIII: Advanced Materials and Technologies*, *2*, 412–554.
18. Breusers, H. N. C., Nicollet, G., & Shen, H. W. (1977). Local scour around cylindrical piers. *Journal of Hydraulic Research*, *15*(3), 211–252.
19. Brunone, B., Karney, B. W., Mecarelli, M., & Ferrante, M. (2000). Velocity profiles and unsteady pipe friction in transient flow. *Journal of Water Resources Planning and Management*, *126*(4), 236–244.
20. Canadell, D., García-Martínez, J., Alepuz, P., Pérez-Ortín, J. E., & Ariño, J. (2015). Impact of high pH stress on yeast gene expression: A comprehensive analysis of mRNA turnover during stress responses. *Biochimica et Biophysica Acta (BBA)-Gene Regulatory Mechanisms*, *1849*(6), 653–664.
21. Carravetta, A., Derakhshan Houreh, S., & Ramos, H. M. (2018). Pumps as turbines. *Springer tracts in mechanical engineering*, 236 (1), 1–18.
22. Carta, J. A., González, J., Cabrera, P., & Subiela, V. J. (2015). Preliminary experimental analysis of a small-scale prototype SWRO desalination plant, designed for continuous adjustment of its energy consumption to the widely varying power generated by a stand-alone wind turbine. *Applied Energy*, *137*, 222–239.

23. Carta, J., Gonzalez, J., & Subiela, V. (2004). The SDAWES project: an ambitious R&D prototype for wind-powered desalination. *Desalination*, *161*(1), 33–48.
24. Cavaleri, L., Alves, J. H., Ardhuin, F., Babanin, A., Banner, M., Belibassakis, K., ... & WISE Group. (2007). Wave modelling–the state of the art. *Progress in Oceanography*, *75*(4), 603–674.
25. Champliaud, H., & Feng, Z. (2018). Numerical and experimental comparisons of pressed blades for large Francis turbine runners manufactured with a reconfigurable pressing setup and a conventional setup. *International Journal of Service and Computing Oriented Manufacturing*, *3*(4), 293–317.
26. Charlier, R. H. (1997). Re-invention or aggorniamento? Tidal power at 30 years. *Renewable and Sustainable Energy Reviews*, *1*(4), 271–289.
27. Charlier, R. H. (2001). Ocean alternative energy: The view from China—'small is beautiful.' *Renewable and Sustainable Energy Reviews*, *5*(4), 403–409.
28. Chen, F. (2010). Kuroshio power plant development plan. *Renewable and Sustainable Energy Reviews*, *14*(9), 2655–2668.
29. Chen, J., Lu, W., Hu, Z., Lei, Y., & Yang, M. (2018). Numerical studies on the performance of a drag-type vertical axis water turbine for water pipeline. *Journal of Renewable and Sustainable Energy*, *10*(4), 044503.
30. Coney, M. W., Linnemann, C., Greenwood, A. L., & Hogg, C. R. (2003, January). First test results of a novel large high-efficiency reciprocating piston engine. In *Internal Combustion Engine Division Spring Technical Conference* (Vol. 36789, pp. 741–750).
31. Coney, M., Linnemann, C., Sugiura, K., & Goto, T. (2003). First Prototype of the High-Efficiency Isoengine. *Institution of Diesel and Gas Turbine Engineers (IDGTE), Power Engineer Journal*, *7*(2).
32. Coughenour, C. L., Archer, A. W., & Lacovara, K. J. (2009). Tides, tidalites, and secular changes in the Earth–Moon system. *Earth-Science Reviews*, *97*(1–4), 59–79.
33. Date, A., Date, A., & Akbarzadeh, A. (2013). Investigating the potential for using a simple water reaction turbine for power production from low head hydro resources. *Energy Conversion and Management*, *66*, 257–270.

34. Davidson, N. C., Laffoley, D. D. A., Doody, J. P., Way, L. S., Gordon, J., Key, R. E., ... & Duff, K. L. (1991). Nature conservation and estuaries in Great Britain. *Nature Conservancy Council, Peterborough*, 1–76.
35. Dong, Z. (2019). Blooms of the moon jellyfish Aurelia: causes, consequences and controls. *World seas: an environmental evaluation*, 163–171.
36. Dursun, B., & Gokcol, C. (2011). The role of hydroelectric power and contribution of small hydropower plants for sustainable development in Turkey. *Renewable Energy*, *36*(4), 1227–1235.
37. Ekama, G. A. (2009). Using bioprocess stoichiometry to build a plant-wide mass balance based steady-state WWTP model. *Water Research*, *43*(8), 2101–2120.
38. Elfarra, M. A., Sezer Uzol, N., & Akmandor, İ. S. (2015). Investigations on blade tip tilting for HAWT rotor blades using CFD. *International Journal of Green Energy*, *12*(2), 125–138.
39. Escaler, X., Egusquiza, E., Farhat, M., Avellan, F., & Coussirat, M. (2006). Detection of cavitation in hydraulic turbines. *Mechanical Systems and signal Processing*, *20*(4), 983–1007.
40. Fabian, A. C., Pringle, J. E., & Rees, M. J. (1975). Tidal capture formation of binary systems and X-ray sources in globular clusters. *Monthly Notices of the Royal Astronomical Society*, *172*(1), 15P-18P.
41. Ferguson, N., Lucks, S., Schneier, B., Whiting, D., Bellare, M., Kohno, T., ... & Walker, J. (2010). The Skein hash function family. *Submission to NIST (Round 3)*, *7*(7.5), 3.
42. Finkl, C. W., & Charlier, R. (2009). Electrical power generation from ocean currents in the Straits of Florida: Some environmental considerations. *Renewable and Sustainable Energy Reviews*, *13*(9), 2597–2604.
43. Flexas, M. D. M., Schodlok, M. P., Padman, L., Menemenlis, D., & Orsi, A. H. (2015). Role of tides on the formation of the A ntarctic S lope F ront at the W eddell-S cotia Confluence. *Journal of Geophysical Research: Oceans*, *120*(5), 3658–3680.
44. Gao, S., Chen, X., Yi, Q., Wang, G., Pan, G., Lin, A., & Peng, G. (2010). A strategy for the proliferation of Ulva prolifera, main causative species of green tides, with formation of sporangia by fragmentation. *PLoS One*, *5*(1), e8571.

45. Ghidaoui, M. S., Zhao, M., McInnis, D. A., & Axworthy, D. H. (2005). A review of water hammer theory and practice. *Appl. Mech. Rev.*, *58*(1), 49–76.
46. Ginzburg, N. S., Zaitsev, N. I., Ilyakov, E. V., Kulagin, I. S., Novozhilova, Y. V., & Sergeev, A. S. (2001). Chaotic generation in a megawatt backward-wave tube. *Technical Physics*, *46*(11), 1420–1427.
47. Haigh, I., Nicholls, R., & Wells, N. (2010). Assessing changes in extreme sea levels: Application to the English Channel, 1900–2006. *Continental Shelf Research*, *30*(9), 1042–1055.
48. Han, L., Duan, X. L., Gong, R. Z., Zhang, G. F., Wang, H. J., & Wei, X. Z. (2019). Physic of secondary flow phenomenon in distributor and bifurcation pipe of Pelton turbine. *Renewable Energy*, *131*, 159–167.
49. Hansson, H. G., Hansson, M. H. G., & Works, K. (1977). Development of the Kaplan Turbine. *Daedalus Stockholm*, 11–35.
50. Herrmann, A. S., Stieglitz, A., Brauner, C., Peters, C., & Schiebel, P. (2013). Polymer matrix composites. *Structural Materials and Processes in Transportation*, 257–302.
51. Horsburgh, K. J., & Wilson, C. (2007). Tide-surge interaction and its role in the distribution of surge residuals in the North Sea. *Journal of Geophysical Research: Oceans*, *112*(C8).
52. Hurley, J. R., Tout, C. A., & Pols, O. R. (2002). Evolution of binary stars and the effect of tides on binary populations. *Monthly Notices of the Royal Astronomical Society*, *329*(4), 897–928.
53. Intriago Zambrano, J. C., Michavila, J., Arenas Pinilla, E., Diehl, J. C., & Ertsen, M. W. (2019). Water lifting water: a comprehensive spatiotemporal review on the hydro-powered water pumping technologies. *Water*, *11*(8), 1677.
54. Jaberg, H., Weber, G., Weber, G., Bauer, C., Gössinger, S., Heninger, L., ... & Kotoul, M. (2015). Kaplan-Turbine. In *Wasserkraftprojekte Band II* 40 (7), 11–75.
55. Jenicek, P., Kutil, J., Benes, O., Todt, V., Zabranska, J., & Dohanyos, M. (2013). Energy self-sufficient sewage wastewater treatment plants: is optimized anaerobic sludge digestion the key?. *Water Science and Technology*, *68*(8), 1739–1744.
56. Kapoor, A., Balasubramanian, S., Kavitha, E., Poonguzhali, E., & Prabhakar, S. (2021). Role of Membranes in Wastewater Treatment. In *Water Pollution and Remediation: Photocatalysis* 34 (3), 247–281.

57. Kenway, S. J., Lant, P. A., Priestley, A., & Daniels, P. (2011). The connection between water and energy in cities: a review. *Water Science and Technology*, *63*(9), 1983–1990.
58. Khan, M. J., Bhuyan, G., Iqbal, M. T., & Quaicoe, J. E. (2009). Hydrokinetic energy conversion systems and assessment of horizontal and vertical axis turbines for river and tidal applications: A technology status review. *Applied Energy*, *86*(10), 1823–1835.
59. Kok, J. K., Scheepers, M. J. J., & Kamphuis, I. G. (2010). Intelligence in electricity networks for embedding renewables and distributed generation. In *Intelligent Infrastructures* 34 (6), 179–209.
60. Kumar, P., & Saini, R. P. (2010). Study of cavitation in hydro turbines—A review. *Renewable and Sustainable Energy Reviews*, *14*(1), 374–383.
61. LaCroix, M. A., Chong, E., Shen, W., Nir, E., Musa, F. A., Mei, H., ... & Tonietto, D. (2021, February). 8.4 A 116Gb/s DSP-Based Wireline Transceiver in 7nm CMOS Achieving 6pJ/b at 45dB Loss in PAM-4/ Duo-PAM-4 and 52dB in PAM-2. In *2021 IEEE International Solid-State Circuits Conference (ISSCC)* (Vol. 64, pp. 132–134).
62. Laghari, J. A., Mokhlis, H., Bakar, A. H. A., & Mohammad, H. (2013). A comprehensive overview of new designs in the hydraulic, electrical equipments and controllers of mini hydro power plants making it cost effective technology. *Renewable and Sustainable Energy Reviews*, *20*, 279–293.
63. Lai, D. (1997). Dynamical tides in rotating binary stars. *The Astrophysical Journal*, *490*(2), 847.
64. Lee, H. M., & Ostriker, J. P. (1986). Cross sections for tidal capture binary formation and stellar merger. *The Astrophysical Journal*, *310*, 176–188.
65. Li, C., Li, L., Li, J., & Zhang, Y. (2017). A sequential-parallel interdependent complement scaling approach with its applications to AP1000 passive containment cooling system. *Nuclear Engineering and Design*, *319*, 149–162.
66. Linnemann, C., & Coney, M. W. (2005). The isoengine: realisation of a high-efficiency power cycle based on isothermal compression. *International Journal of Energy Technology and Policy*, *3*(1–2), 66–84.
67. Linnemann, C., Coney, M. W., & Price, A. (2003, January). The Isoengine: A Novel High Efficiency Engine With Optional Compressed

Air Energy Storage (CAES). In *International Joint Power Generation Conference* (Vol. 36924, pp. 873–880).

68. Liu, D., Liu, S., Wu, Y., & Liu, X. B. (2009). Numerical simulation of hydraulic turbine based on fluid-structure coupling. In *Fluid Machinery and Fluid Mechanics* 213, 345–351.

69. Liu, X., Luo, Y., Karney, B. W., & Wang, W. (2015). A selected literature review of efficiency improvements in hydraulic turbines. *Renewable and Sustainable Energy Reviews*, *51*, 18–28.

70. Mack, R., Drtina, P., & Lang, E. (1999). Numerical prediction of erosion on guide vanes and in labyrinth seals in hydraulic turbines. *Wear*, *233*, 685–691.

71. Majewski, P. W., & Yager, K. G. (2016). Rapid ordering of block copolymer thin films. *Journal of Physics: Condensed Matter*, *28*(40), 403002.

72. Marcus, F., Howard, T., Bishop, J., & Director, C. D. (2016). TO: The State Water Resources Control Board. *Cell*, *415*, 0419.

73. Marongiu, J. C., Leboeuf, F., Caro, J., & Parkinson, E. (2010). Free surface flows simulations in Pelton turbines using an hybrid SPH-ALE method. *Journal of Hydraulic Research*, *48*(sup1), 40–49.

74. McMillan, S. L., McDermott, P. N., & Taam, R. E. (1987). The formation and evolution of tidal binary systems. *The Astrophysical Journal*, *318*, 261–277.

75. Mishra, R. K., Thomas, J., Srinivasan, K., Nandi, V., & Bhatt, R. R. (2017). Failure analysis of an un-cooled turbine blade in an aero gas turbine engine. *Engineering Failure Analysis*, *79*, 836–844.

76. Mishra, S., Singal, S. K., & Khatod, D. K. (2012). A review on electromechanical equipment applicable to small hydropower plants. *International journal of energy research*, *36*(5), 553–571.

77. Momčilović, D., & Odanović, Z. (2019). Feasibility Analysis for Repair of Large Diameter Hydro Turbine Shaft Based on Weldability Test and Welding Cycle Simulation. In *Advanced Materials Research* (Vol. 1153, pp. 36–45).

78. Momčilović, D., Odanović, Z., Mitrović, R., Atanasovska, I., & Vuherer, T. (2012). Failure analysis of hydraulic turbine shaft. *Engineering Failure Analysis*, *20*, 54–66.

79. Morabito, A., e Silva, G. D. O., & Hendrick, P. (2019). Deriaz pump-turbine for pumped hydro energy storage and micro applications. *Journal of Energy Storage*, *24*, 100788.
80. Nasir, B. A. (2014). Suitable selection of components for the micro-hydro-electric power plant. *Advances in energy and power*, *2*(1), 7–12.
81. Nihous, G. C. (2010). Mapping available Ocean Thermal Energy Conversion resources around the main Hawaiian Islands with state-of-the-art tools. *Journal of Renewable and Sustainable Energy*, *2*(4), 043104.
82. Nowak, O., Keil, S., & Fimml, C. (2011). Examples of energy self-sufficient municipal nutrient removal plants. *Water Science and Technology*, *64*(1), 1–6.
83. Pawlowicz, R., Beardsley, B., & Lentz, S. (2002). Classical tidal harmonic analysis including error estimates in MATLAB using T_TIDE. *Computers & Geosciences*, *28*(8), 929–937.
84. Pezzinga, G. (2000). Evaluation of unsteady flow resistances by quasi-2D or 1D models. *Journal of Hydraulic Engineering*, *126*(10), 778–785.
85. Pingree, R. D. (1978). The formation of the Shambles and other banks by tidal stirring of the seas. *Journal of the Marine Biological Association of the United Kingdom*, *58*(1), 211–226.
86. Pingree, R. D., & Maddock, L. (1979). The tidal physics of headland flows and offshore tidal bank formation. *Marine Geology*, *32*(3–4), 269–289.
87. Puertas, I., Pérez, C. L., Salcedo, D., León, J., Fuertes, J. P., & Luri, R. (2013). Design and mechanical property analysis of AA1050 turbine blades manufactured by equal channel angular extrusion and isothermal forging. *Materials & Design (1980–2015)*, *52*, 774–784.
88. Putnam, A. A., Belles, F. E., & Kentfield, J. A. C. (1986). Pulse combustion. *Progress in energy and combustion science*, *12*(1), 43–79.
89. Quoilin, S., Van Den Broek, M., Declaye, S., Dewallef, P., & Lemort, V. (2013). Techno-economic survey of Organic Rankine Cycle (ORC) systems. *Renewable and Sustainable Energy Reviews*, *22*, 168–186.
90. Rauf, O., Wang, S., Yuan, P., & Tan, J. (2015). An overview of energy status and development in Pakistan. *Renewable and Sustainable Energy Reviews*, *48*, 892–931.

91. Ray, A., Kembhavi, A. K., & Antia, H. M. (1987). Evolution of stellar binaries formed by tidal capture. *Astronomy and Astrophysics*, *184*(1–2), 164–172.

92. Remy, C., Boulestreau, M., & Lesjean, B. (2014). Proof of concept for a new energy-positive wastewater treatment scheme. *Water Science and Technology*, *70*(10), 1709–1716.

93. Rielly, C. D., & Gimbun, J. (2009). Computational fluid mixing. *Food Mixing: Principles and Applications. Chichester, Wiley Blackwell, UK*, 125–174.

94. Rojstaczer, S., & Agnew, D. C. (1989). The influence of formation material properties on the response of water levels in wells to Earth tides and atmospheric loading. *Journal of Geophysical Research: Solid Earth*, *94*(B9), 12403–12411.

95. Salter, S. H., Taylor, J. R. M., & Caldwell, N. J. (2002). Power conversion mechanisms for wave energy. *Proceedings of the Institution of Mechanical Engineers, Part M: Journal of Engineering for the Maritime Environment*, *216*(1), 1–27.

96. Shallcross, D. C. (2013). Safety education through case study presentations. *Education for chemical engineers*, *8*(1), e12-e30.

97. Sharma, H., & Singh, J. (2013). Run off river plant: status and prospects. *International Journal of Innovative Technology and Exploring Engineering (IJITEE)*, *3*(2), 2278–3075.

98. Shrestha, U., & Choi, Y. D. (2020). A CFD-Based Shape Design Optimization Process of Fixed Flow Passages in a Francis Hydro Turbine. *Processes*, *8*(11), 1392.

99. Siddiqi, A., & Anadon, L. D. (2011). The water–energy nexus in Middle East and North Africa. *Energy policy*, *39*(8), 4529–4540.

100. Singh, P., & Nestmann, F. (2009). Experimental optimization of a free vortex propeller runner for micro hydro application. *Experimental Thermal and Fluid Science*, *33*(6), 991–1002.

101. Sockel, H. (2014). Essential Contributions of Austria to Fluid Dynamics Prior to the End of World War II. In *The History of Theoretical, Material and Computational Mechanics-Mathematics Meets Mechanics and Engineering* 50 (3), 355–382.

102. Solemslie, B. W., & Dahlhaug, O. G. (2012, September). A reference Pelton turbine design. In *IOP Conference Series: Earth And Environmental Science* (Vol. 15, No. 3, p. 032005).

103. Stillwell, A. S., King, C. W., Webber, M. E., Duncan, I. J., & Hardberger, A. (2011). The energy-water nexus in Texas. *Ecology and Society*, *16*(1).
104. Svardal, K., & Kroiss, H. (2011). Energy requirements for waste water treatment. *Water Science and Technology*, *64*(6), 1355–1361.
105. Tai, W., Carley, L. R., & Ricketts, D. S. (2013, February). A 0.7 W fully integrated 42GHz power amplifier with 10% PAE in 0.13 μm SiGe BiCMOS. In *2013 IEEE International Solid-State Circuits Conference Digest of Technical Papers* 48(4), 142–143.
106. Tiwari, G., Kumar, J., Prasad, V., & Patel, V. K. (2020). Utility of CFD in the design and performance analysis of hydraulic turbines—A review. *Energy Reports*, *6*, 2410–2429.
107. Viollet, P. L. (2017). From the water wheel to turbines and hydroelectricity. Technological evolution and revolutions. *Comptes Rendus Mécanique*, *345*(8), 570–580.
108. Von Jouanne, A., & Brekken, T. K. (2019). Ocean and geothermal renewable energy systems. *Power Electronics in Renewable Energy Systems and Smart Grid: Technology and Applications*, 237 (7), 391–441.
109. Wagner, H. J., & Mathur, J. (2011). Hydraulic Turbines: Types and Operational Aspects. In *Introduction to Hydro Energy Systems* 123 (1), 71–93.
110. Waters, S., & Aggidis, G. (2016). Tidal range technologies and state of the art in review. *Renewable and Sustainable Energy Reviews*, *59*, 514–529.
111. Wei, W., Zhou, X., Wang, D., Sun, J., & Wang, Q. (2017). Free ammonia pre-treatment of secondary sludge significantly increases anaerobic methane production. *Water research*, *118*, 12–19.
112. Woldemariam, E. T., Lemu, H. G., & Wang, G. G. (2018). CFD-Driven Valve Shape Optimization for Performance Improvement of a Micro Cross-Flow Turbine. *Energies*, *11*(1), 248.
113. Wong, Y. W., & Sumathy, K. (1999). Solar thermal water pumping systems: a review. *Renewable and Sustainable Energy Reviews*, *3*(2–3), 185–217.
114. Woodworth, P. L. (2010). A survey of recent changes in the main components of the ocean tide. *Continental Shelf Research*, *30*(15), 1680–1691.

115. Woodworth, P. L., & Blackman, D. L. (2004). Evidence for systematic changes in extreme high waters since the mid-1970s. *Journal of Climate*, *17*(6), 1190–1197.

116. Woodworth, P. L., Teferle, F. N., Bingley, R. M., Shennan, I., & Williams, S. D. P. (2009). Trends in UK mean sea level revisited. *Geophysical Journal International*, *176*(1), 19–30.

117. Yuksek, O., Komurcu, M. I., Yuksel, I., & Kaygusuz, K. (2006). The role of hydropower in meeting Turkey's electric energy demand. *Energy Policy*, *34*(17), 3093–3103.

INDEX

F

G

H

I

K

M

N

O

P

R